Emergency Response to Domestic Terrorism

EMERGENCY RESPONSE TO DOMESTIC TERRORISM

How Bureaucracies Reacted to the 1995 Oklahoma City Bombing

by
Alethia H. Cook

NEW YORK • LONDON

2009

The Continuum International Publishing Group Inc
80 Maiden Lane, New York, NY 10038

The Continuum International Publishing Group Ltd
The Tower Building, 11 York Road, London SE1 7NX

www.continuumbooks.com

Copyright © 2009 by Alethia H. Cook

All rights reserved. No part of this book may be reproduced, stored in a retrieval system, or transmitted, in any form or by any means, electronic, mechanical, photocopying, recording, or otherwise, without the written permission of the publishers.

Library of Congress Cataloging-in-Publication Data
A catalog record for this book is available from the Library of Congress.

ISBN: 978-0-8264-3073-1

To Akron, OH FD Lt. Keith Hillman (Ret.); Bethel Township, OH FD Lt. Craig Hillman; Akron, OH FD Lt. Bob Hillman (Ret.); and USAF Maj. Vanessa Hillman (Dad, Brother, Uncle, and Sister respectively) and all other emergency responders and Service men and women who are heroes in my eyes. They will tell you that they are just doing their jobs when they run toward danger (as I was told numerous times in my interviews), so they are not heroes. I find them heroic for choosing careers that so frequently put them in danger.

Contents

List of Figures and Tables   viii

Acknowledgments   ix

Abbreviations   x

Chapter One Bureaucratic Response to Disasters: Issues and Methods   1

Chapter Two Disaster, Chaos, and Response: First Arrival at the Murrah Building Scene   20

Chapter Three Emergency Response Challenges   33

Chapter Four Response as a Street-Level Phenomenon   51

Chapter Five Response Bureaucracies' Tasks and Goals   66

Chapter Six Conclusions: Lessons Learned and Reinforced   83

Appendix A: Interviews Conducted   101

Appendix B: Interview Questionnaires   103

Bibliography   115

Index   123

List of Figures and Tables

Figures

2.1	Layout of the Scene	22
2.2	Blast's Impacts on the Murrah Building	24
2.3	Extensive Damage Caused to Surrounding Buildings	25
4.1	Basic ICS Structures	54
4.2	Partial Representation of OCFD Murrah Incident Command	56

Tables

2.1	Timeline of Events, Oklahoma City, April 19, 1995, 9:02 a.m.–9:00 p.m.	28
5.1	Key Tasks Performed by Participating Response Organizations Based on Interviews and AAR	66
6.1	Summary of Comment Frequencies	84
6.2	Minimal Representation at Major Multiagency Training Exercises	84
6.3	Evaluation of ESFs, Oklahoma City Emergency Operations Plans, and Response Actions Taken on April 19, 1995 in Oklahoma City (Key and Sources)	86

Acknowledgments

When I started this project I had no idea how many people would give so much of themselves to help me complete it. I would be remiss if I did not start by giving considerable thanks to the participants in the Oklahoma City bombing response who allowed me to interview them. Their participation made this project possible. I am aware that talking to me was difficult for many of them and that some have been interviewed repeatedly since the incident. I deeply appreciate their willingness to revisit that day to assist me in this project. The Memorial Institute for the Prevention of Terrorism (MIPT) was also crucial to the success of this project. The MIPT's assistance in arranging my interviews and access to the archives was invaluable. In particular, I would like to thank Brad Robison, Michael Vetti, Jane Thomas, and Ken Thompson who all worked so hard to make my visit to Oklahoma City and this book successful.

I must also recognize David Louscher for his contribution to my scholarly development in general and this project in particular. He has been a mentor and a friend to me for a very long time. Working with him has improved my analytical and writing skills to the level represented in this book.

A major contribution to this project was made by the academic community that has surrounded me for most of my life. Of particular assistance in this project were Steve Hook and Mark Cassell. They pushed me to make this project stronger, more focused, and more scholarly. Their insights and patience are truly appreciated. Thanks also to Thomas Hensley and Thomas Schmidlin for their participation in the process.

To my family, who provided persistent encouragement and celebrated each milestone with me, I thank you. My family's faith that I would finish this project never wavered, even when mine had.

Finally, I would like to say thank you to the emergency responders in my family: my father (Keith Hillman), brother (Craig Hillman), and uncle (Bob Hillman). Doing these interviews and the research surrounding them made me understand more fully the risks you each have taken to help others. I appreciate the work that you have done to keep our communities safe and am proud of your service.

Abbreviations

AAR	After-Action Report
BATF	Bureau of Alcohol, Tobacco, and Firearms
CISD	Critical Incident Stress Debriefings
COWS	Cellular-on-Wheels
D-MORT	Disaster Mortuary
DHS	Department of Homeland Security
EMS	Emergency Medical Services
EMSA	Emergency Medical Services Authority
EOC	Emergency Operations Center
ESF	Emergency Support Function
FAC	Family Assistance Center
FBI	Federal Bureau of Investigations
FEMA	Federal Emergency Management Agency
GSA	General Services Agency
ICS	Incident Command System
MIPT	Memorial Institute for the Prevention of Terrorism
NIMS	National Incident Management System
NRF	National Response Framework
NRP	National Response Plan
OCFD	Oklahoma City Fire Department
OCPD	Oklahoma City Police Department
ODCEM	Oklahoma Department of Civil Emergency Management
PIO	Public Information Officer
TAFB	Tinker Air Force Base
TIA	Terrorist Incident Appendix
USAR	Urban Search and Rescue

CHAPTER ONE

Bureaucratic Response to Disasters: Issues and Methods

Introduction

The Oklahoma City bombing response is an example of a challenging and complicated bureaucratic response to an unexpected, unprecedented, and chaotic act of terrorism. The bomb was detonated at 9:02 a.m. on April 19, 1995. Starting at that moment and progressing over twenty-one days, the response to the bombing was a major undertaking for the community and its responders. While over ten years have passed since this incident, there have been few attempts, outside the Oklahoma City response community, to analyze the response systematically and learn lessons from what occurred.

One way to learn more about effective response is to undertake an intensive study of past response efforts. Through a disciplined and structured evaluation, such a study could provide insights into the following: the key challenges faced by the responders; the impacts of bureaucratic structures, networks, and culture; and the successful and unsuccessful aspects of the response efforts. This type of analysis can also provide more information about how emergency response activities might be evaluated. Furthermore, the research will provide emergency response planners with valuable insights into the challenges faced when street-level bureaucrats respond to terrorism or other major disasters. This study will provide increased insights into the strengths and weaknesses of emergency response practices. The lessons learned will benefit attempts to improve emergency response to both terrorist attacks and natural disasters.

This work has important ramifications beyond its identification of lessons learned and reinforcement of an already extensive literature on emergency response. The findings are important to any society that might one day experience disaster. This work draws on literature in political science, public administration, public policy, sociology, and emergency management to arrive at conclusions about disaster response. Drawing from multiple disciplines allows for a more comprehensive understanding of the response to the incident as well as making the analysis richer.

A community's foundations and their faith in their government could be undermined by poor response. The ability of a city to continue to function successfully after a disaster is tied to the response that is carried out and the successes and failures of the response community. The lessons learned provide advice to local emergency response planners that may help them and their citizens to overcome the impact of terrible events.

Context

Natural disasters and accidents have been persistent challenges faced by the emergency response community. Every incident brings about unique problems that must be solved based on the specific type of incident and the impact it had on the community. The job of first responders became more difficult when the threat of a terrorist attack on the U.S. homeland increased in the 1990s. Commonalities and differences between the challenges posed by terrorist incidents as compared to natural disasters and accidents are discussed later in this work. What is important to recognize here is that terrorist attacks constitute the addition of a major category of disaster to the types of incidents for which governments must prepare which was not always a priority for governments as they made response plans. For that reason, and due to the fact that the case under consideration here was brought about by an act of terrorism, this section provides a discussion of the changing nature of terrorism in the 1990s to provide some context.

In the 1990s, the threat perception about terrorism against American targets changed for attentive scholars and members of the U.S. government. Several key events brought about this reconsideration of the nature and extent of the threat. The first major incident was the 1993 bombing of the World Trade Center, which demonstrated that international terrorist organizations could and would strike at American targets within the continental United States. While in general the American public tends to treat this incident as a rather minor one, it was truly significant both in that it demonstrated this will to harm Americans in the homeland and in the localized economic and physical impacts it inflicted on individuals and firms within the building. Two major incidents occurred in 1995. In March of that year, the Aum Shinrikyo doomsday cult conducted a sarin nerve gas attack on the Tokyo subway, which proved that a strike with a weapon of mass destruction was not beyond the abilities of a terrorist organization and that they may have the desire to use them. This was followed by the April bombing of the Alfred P. Murrah Building in Oklahoma City, which showed that domestic terrorist groups had both the will and means to cause serious destruction. In 1998, terrorists demonstrated sophisticated planning, organizational, and execution capabilities by carrying out nearly simultaneous attacks on U.S. Embassies in Tanzania and Kenya in 1998, killing over 200 people (12 Americans). The 1990s also brought a general shift in the goals of terrorist acts. Before this time, terrorist groups sought sympathetic media coverage. According to noted terrorism scholar Brian Jenkins, the terrorists wanted a lot of people watching, not a lot of people dead.[1] During the 1990s, terrorist methods shifted toward more violent acts that were specifically planned to cause large numbers of casualties.[2]

These changes went unnoticed by most Americans. In a 1999 public opinion survey by the Chicago Council on Foreign Relations, terrorism was not listed by either the general public or by elites as one of the top three problems facing the country.[3] This finding is surprising given the pivotal terrorist events of the 1990s. In 2003, after the terrorist acts of 9/11, it was not at all surprising that 36 percent of respondents to the quadrennial survey spontaneously listed terrorism as one of the top problems facing the nation.[4]

In spite of this lack of public interest pre-9/11, some members of government became more active in seeking a higher level of preparation to respond to a serious terrorist strike against the United States. Several major studies were done by the government and by scholars that indicated the United States was not prepared to respond to an attack, especially if it involved weapons of mass destruction (WMD).[5] Government reacted to the terrorism of the 1990s with laws and programs with the goal of improving response.[6] However, while these actions served to increase government's attention to terrorism and increased the national budget for terrorism response initiatives, no efforts were undertaken to address preparedness shortcomings in a systematic or coherent fashion nationwide.

Terrorism is a complex threat. Many potential terrorist groups are active internationally and within the United States that may wish to bring about change or increase their visibility, and thereby the visibility of their cause, through violence. Furthermore, many potential methods exist through which terrorists may try to bring about instability or change. Conventional explosives have been the most prevalent mode of mass terrorism.[7] However, unconventional terrorist strikes, which may include chemical, biological, nuclear, or radiological attacks, have become increasingly likely. This fact was demonstrated by the sarin nerve gas on the Tokyo subway in 1995 and anthrax attacks in the United States in 2001. The final complicating factor of terrorism is the multiplicity of domestic targets among which terrorists may choose. Taken together, these factors make it unlikely that the United States will be able to deter or intervene effectively in every terrorist strike. For that reason, the United States must become better prepared to respond to attacks so that we may minimize the impact of these events.

Before 9/11, many communities failed to prepare adequately for a potential terrorist attack. Major disasters are low probability/high consequence events characterized by uncertainty.[8] A major terrorist strike would certainly fit into this category. Planning to be prepared for such an incident is difficult because, "people are unlikely to give priority of attention to an unlikely future disaster when there are fifteen tasks that have to be accomplished by Friday. This factor is particularly salient in contemporary government where there are so many programs competing for scarce resources."[9] Furthermore, the public's concern about terrorism tends to wane over time. As stated by emergency response scholar Erik Auf der Heide, "interest in disaster preparedness is proportional to the recency [sic] and magnitude of the last disaster."[10] The transience of public interest makes it difficult for decision-makers to devote significant time and budgetary resources to emergency response planning over time. There is also a tendency for some, particularly smaller, localities to believe that terrorism is unlikely to happen in their community.

There is some disagreement in the literature about whether major events provide the opportunity to learn lessons that will improve future response activities. Some sources argue that crises provide opportunities for emergency response planning to occur.[11] While it is true that there is ample material from which to learn after these events, it is not clear that lessons are indeed learned. Thomas Birkland argues that "the paradox of learning driven by truly large events is that whereas such events provide significant fodder for learning, they are also likely to overwhelm the ability of a system to respond with routine procedures and therefore may limit learning."[12] Birkland also argues that, while lessons are there, they are only useful if they are acted on by elites. Instead of this happening, however, policies generally proceed without reference to the After-Action Reports (AARs) or lessons learned generated by the relevant agencies.[13] Evidence of this is found in the fact that, in spite of 50 years of attempts to improve emergency response, the Hurricane Katrina response was a failure on many levels.[14] In many ways, the magnitude of major incidents may serve as a hindrance to learning, if for no other reason than some may believe that the uniqueness of each response and infrequence of major disasters minimizes the potential inferences that can be drawn for other instances.

Emergency Response Concepts

Preparing for a major incident response is a process that goes on at all levels of government. While response is fundamentally a local-level occurrence, maintaining local support for response planning and funding can be difficult. It is easier, by comparison, to stimulate interest in response budgets and plans at the national level because, "from a national perspective, disasters occur more often and create higher total damages and costs. So it is not surprising that the push for emergency planning comes from the top down: from national to regional to local levels."[15]

However, emergency response planning impacts a community much broader than its practitioners and policy-makers. As emphasized by Brian Toft and Simon Reynolds, "disasters are socio-technical in nature as opposed to solely, or even mainly, technical."[16] Every citizen is impacted by the emergency response activities of relevant agencies. The entire society is altered through the experience. No one would argue that the only impact of Hurricane Katrina was on the emergency responders of New Orleans. Clearly, the societal impact there was profound. The emergency response activities can also impact citizens' support for specific members of government as they seek reappointment or reelection.[17] While the focus of this book is on the more technical aspects of emergency response, it is important to emphasize the broader social and political implications of response activities. In the end, minimizing the social impacts of disaster is the fundamental reason for the existence of emergency response entities.

It is important at this time to discuss the distinctions among various aspects of emergency response and the terminology that is being used in this text. While it would be an exaggeration to say that the process for emergency activities is organized in truly discrete phases, it is perhaps best to approach it from that perspective yet recognize that the discussion oversimplifies the process. There are many different conceptualizations of these phases. This one represents the integration of several authors' work. It is useful to think of these activities as those that occur before, during, and after the incident.[18]

Before the incident occurs preparations for response have taken place. The preparations may include such activities as provision of physical security for specific locations, developing communications networks, and other activities. Being prepared for response helps communities to absorb the impacts of incidents better and to recover from them more quickly. Planning is a component of preparation. American government at all levels engages in planning activities for emergency response.[19] These activities are currently guided by the National Response Framework (NRF) (and the preceding Federal Response Plan (FRP) and National Response Plan (NRP)), which identifies those agencies with primary responsibility for specific types of response activities and those that will provide support to the lead agency.[20]

Emergency response occurs as the incident evolves and immediately after it has taken place.[21] This involves addressing the impacts of an incident in its immediate aftermath. Depending on the nature of the incident, it may include such activities as putting out fires, assisting the injured, securing the scene, and much more. After this initial phase of the response (which could certainly range in duration from hours to days, depending on the magnitude of the incident), the activities would shift to recovery, which focuses on dealing with the longer-term impacts of the incident.[22] Recovery includes the community's actions to rebuild damaged infrastructure, resettle individuals displaced by the incident, return to normal operations, and other similar activities. This book contains information relevant to all phases of the process, with the exception of recovery, which was outside the scope of the interests of the study.

Due to the crucial activities of the early stages of the response, this book focuses on the first 12 hours of the response. This is a period of time during which the local responders had limited assistance in dealing with the tragedy from federal authorities. Response activities undertaken during the first 12 hours have the greatest potential to save additional lives and minimize the impact of the disaster. For this reason, the focus was restricted primarily to preparedness and response. A more thorough explanation for the focus on the first 12 hours of the response is provided in the methods section within this chapter.

The perception of effective response is, in many ways, dependent on each on these phases proceeding smoothly. The nature of American society and economy make it important that emergency response be carried out effectively. As argued by Louise Comfort, "cities represent a major investment of not just local funds, but also substantial investments by the region, state and the nation."[23] Thus, there is an economic impact from a major disaster in excess of the losses to the locality. More importantly, however, the social and political impacts of the response activities reverberate throughout the country. Citizens' impressions of government performance at all levels are influenced by the efficacy of the aid given at the local level. Based on the preceding discussion, successful response will be defined as one that limits additional casualties and damage, while minimizing the impact of the incident on the surrounding community.

Identified Response Successes and Challenges

As can be expected, through the course of the initial phases of the response to the Murrah Building bombing some things went well while others did not. Among the more successful aspects of the response were as follows:

Local-level networking facilitated multiagency response

The Oklahoma City response agencies cooperated effectively to initiate response activities.[24] In many ways, this cooperation was facilitated by the regular training and networking that took place before the incident within the community. Without such cooperation, the chaotic scene could have been exacerbated. In spite of the high level of cooperation of the response community the situation was still chaotic, but the responders were able to work together to bring order to the scene. In addition, the following are other outcomes that were facilitated by training and networking:

Rapid mobilization expedited response and probably saved lives

The response to the disaster unfolded almost immediately after the incident occurred. The blast occurred at 9:02 a.m. Responders started arriving on the scene within minutes, and a command post was in place by 9:08 a.m.[25] Rapid response is generally seen as the best means for mitigating the effects of a disaster.[26]

The injured were rescued and treated quickly

Similarly, the majority of the live victims of the attack were removed from the building and taken to hospitals relatively quickly.[27] The rapid movement of the injured to triage stations and then medical facilities likely saved the lives of some victims and helped to minimize the impact of the attack.

Victims' families received support

Responders also recognized early in the response that they would have to find ways to address the needs of the families of the victims. Redirecting families from the scene and developing an area where they could wait for news about their loved ones occurred early in the response process and benefited the responders by removing that factor from the scene.[28]

In spite of regular planning and training activities, there will likely always be aspects of responses to major disasters that do not go well. The unique aspects of each incident as well as common challenges experienced in such responses can be expected to stymie the response of even well-trained individuals. In this case, the less effective aspects of the response included the following:

Bureaucratic cultures clashed

This was perhaps because of a lack of networking (across the levels of government and among the national-level response entities), resulting in federal assistance being integrated slowly. While coordination among local entities was relatively effective, incorporation of federal entities into the local response was more difficult. This was largely attributed by those interviewed to the fact that the Federal Bureau of Investigations (FBI) and Federal Emergency Management Agency (FEMA) had not yet had to coordinate the priorities of search and rescue (FEMA) with the requirements for managing a crime scene (FBI).[29] Local responders who were interviewed also felt unsure of the role that FEMA would play, with some feeling that FEMA wanted to take over.[30]

Media, volunteers, and donations were dealt with ineffectively

A lack of training for specific aspects of the response resulted in donations, volunteers, and the media being dealt with inefficiently. Because this was the first major incident of this type in the United States, the local response community did not appreciate the intensity with which the people of Oklahoma City and the rest of the country would respond to the incident. The outpouring of volunteers and donations overwhelmed responders.[31] Figuring out the logistics of using, staging or storing the assistance took a great deal of the responders' time in the initial stages of the disaster. Similarly, the local response community was challenged in its dealing with the massive influx of media outlets, which arrived on the scene from all over the world to cover the story.

Freelancing and volunteers complicated response

Response was made difficult by the influx of volunteers and "freelancing" responders—both of which functioned outside of bureaucratic standard operating procedures. Securing the scene was hampered by the responders' determination to rescue the injured. Instead of focusing on setting safe perimeters, many of the first responders on the scene provided emergency first aid and entered buildings searching for victims.[32] This lack of discipline complicated the response by letting more citizens onto the scene and making it impossible to account for those who were there. There are strong arguments to be made that this type of extraorganizational intervention saves lives. However, emergency response plans and training emphasize discipline among the responders to maintain accountability and the safety of the responders themselves.

Despite the time that has passed since this 1995 incident, there has been relatively little change in the challenges and the means available to address them. Many of the difficulties experienced in the Oklahoma City response (communicating among responders, coordinating activities across levels of government, dealing with an influx of supplies and volunteers, etc.) were repeated on 9/11, during the response to Hurricane Katrina, and again during Hurricane Ike. The persistence of these challenges across incidents of different natures and through the passing of time reinforces Birkland's argument that lessons about response to major incidents remain to be learned.

Improving emergency response preparedness to acts of terrorism would have the added benefit of increasing capabilities to respond to natural disasters. Hurricane Katrina demonstrated that the country's emergency response preparations continue to be lacking. Like a major terrorist attack, a hurricane the scale of Katrina would be deemed a low probability/high impact event. The Katrina response was plagued with significant difficulties including a failure in communications technologies, a lack of adequate supplies, the need to shelter large numbers of citizens, insufficient government coordination at each level of government and across the levels of government, and a large number of injured and dead.[33] These characteristics would also be likely to be present in a major terrorist attack.

Terrorism response is arguably more complex than responding to a naturally occurring event or an accident in that the responders must be aware of the fact that their crisis scene is also a crime scene. (This is an issue that will be discussed in greater detail later in the book.) The need for coordination sets the stage for significant conflict between law enforcement's desire to protect evidence for the prosecution of the perpetrators and the other emergency response elements that may be destroying evidence in carrying out their response efforts.

The United States must become better able to learn from the challenges encountered in past disaster response activities if it is to improve emergency response to acts of terrorism. While there are significant differences between a terrorist incident and a natural disaster, there are consistent problems that have been experienced in disaster responses that will be present regardless of the cause of the disaster. According to Erik Auf der Heide an examination of disaster responses and their difficulties shows that there is not much being learned. He argues that not only are communities not learning from each others' experiences, but individual communities are also not learning from and correcting their own shortfalls from previous responses.[34]

However, in spite of the post-9/11 increased focus on the threat of terrorism, "comparatively little attention is paid to the need for a more comprehensive response, including resolving the crisis if it is continuing, reducing the impact of the violence on the targets and other audiences, reducing the danger to public health and safety, and providing immediate care to the injured."[35] Furthermore, not enough attention has been paid to response planning and response policy evaluation. This research project provides a useful methodology for evaluating response effectiveness.

Key Arguments

The theoretical literature regarding bureaucratic functions and performance has increased the understanding of how public policy is implemented. Through this literature, insights have been gained into how bureaucracies operate and under what conditions they can be successful. This book draws extensively upon that literature and presents the case of the Oklahoma City bombing response with a focus on three key bureaucratic characteristics: structures, networks, and culture. While there are many other bureaucratic concepts that are clearly relevant to response, these three were selected due to their strong overlap with the emergency response literature and their important impacts on the relative success of response activities.

In terms of structures, the book focuses on differing conceptualizations of policy implementation in the American system of federalism. Four major management models include the top-down model, the donor-recipient model, the jurisdiction-based model, and the network model.[36] The top-down theoretical argument depicts a management arrangement that is highly centralized with a strong system of federal control.[37] Under this arrangement, the local-level bureaucratic entities exhibit high levels of compliance to federal decisions. In contrast, the donor-recipient model depicts a more collaborative system between the federal and local entities. Donor-recipient management is "based on mutually dependent actors in the intergovernmental system rather than complete federal control and influence."[38] This type of management exhibits a less hierarchical system with more of a bottom-up administration for certain policies.[39] In fact, there are arguments that some "policy problems may require—by virtue of their technical requirements, if not by the mandate itself—substantial local presence."[40] In the jurisdiction-based model of management, the local officials are primary drivers for policy implementation and decision-making. This involves local managers developing relationships to pursue strategically the interests of their area.[41] Finally, in the network model—which is distinct from theoretical arguments about the presence and nature of policy networks that will be discussed later—there is collaborative leadership among a large number of local entities, "none of which possesses the power to determine the strategies of other actors."[42] Decision-making in this type of model is characterized by interdependency, where there are no central actors driving the process.[43] The case study will examine when and under what conditions the emergency management of the incident response resembled these different management styles.

In his authoritative work on bureaucracy, James Q. Wilson explained that many scholars examine bureaucracies from the top down, exploring the "structure, purposes, and resources of the organization."[44] Strictly using this top-down approach is inadequate for the examination of terrorism response, as responders' decisions at the bottom of the bureaucratic hierarchy will have significant impact on the policies (in some cases changing them radically) as they attempt to find the most effective methods of response to a given situation. As mentioned above, first responders exhibit several characteristics of bureaucracies as put forth by Wilson. In particular, among emergency responders, "there is no distinction between 'policy' and 'administration'; almost every administrative act has policy implications and may, indeed, *be* policy whether intended or not."[45] In a similar argument, B. Guy Peters claimed that "to a great extent the 'real' policy of government is that policy which is implemented rather than that policy which is adopted by the legislature of the upper echelons of the bureaucracy."[46]

These arguments are particularly relevant to this discussion of emergency response to acts of terrorism. Crises, such as the need to respond to a terrorist act, are extreme cases that challenge bureaucratic agencies to react. In many ways, normal bureaucratic behavior is disrupted by crisis events. In particular, in a crisis situation standard operating procedures (i.e. policies) tend to "give way to informal processes and improvisation."[47] This phenomenon is not restricted to first response. Rather, it can be observed across a wide variety of different policy areas. In fact, Jeffrey Pressman and Aaron Wildavsky urged scholars to merge analyses of policy making and implementation into one another because the improvisation that takes place during implementation inherently alters the nature of policy.[48]

Wilson argued that a bottom-up approach may be more appropriate to certain research tasks. "By looking at bureaucracies from the bottom up, we can assess the extent to which their management systems and administrative arrangements are well or poorly suited to the tasks the agencies actually perform."[49] In fact, some scholars take Wilson's argument further, stating that some policy problems may require a bottom-up approach due to the need for a substantial local presence in the issue area or because implementation has to be flexible.[50] According to Wilson, such an approach necessitates that the researcher understands how bureaucratic organizations decide how to perform their critical tasks, defined by Wilson as "those behaviors which, if successfully performed by key organizational members, would enable the organization to manage its critical environmental problem."[51] In the case of the Oklahoma City bombing, the critical environmental problem facing emergency response bureaucracies was the aftermath of a terrorist strike. Those critical environmental problems will be different for each response bureaucracy. For instance, the fire department may be concerned with putting out the fire, while police forces may be engaged in setting up a safe perimeter.

The bottom-up approach requires that the planning and actions taken at the lowest levels, represented in this work by the emergency responders, be understood. Relevant works include those that explore the interface between so-called street-level bureaucrats and both the public and the upper levels of bureaucracy. Michael Lipsky defines street-level bureaucrats as "public service workers who interact directly with citizens in the course of their jobs and who have substantial discretion in the execution of their work."[52] Key arguments within this literature deal with the resource constraints faced by street-level bureaucrats and the tasks and goals of street-level bureaucrats.

The second driving concept being evaluated by this book is the role and form of policy networks that may be functioning in the area of emergency response. Specifically, this aspect of the book seeks to explore the nature of the relationships among the bureaucratic organizations, legislators, and private sector actors such as lobbyists. Theoretically, there are a variety of different interrelationships that have been hypothesized to exist among these types of actors. The so-called Iron Triangle is a conceptualization of the relationships that is sometimes also referred to as a subgovernment relationship. This type of network would be characterized by the mutually beneficial trade-offs of policy, administration, and support that occur among the participants. These types of networks are typically highly formalized, fixed and stable, with participants being consistent.[53] Issue networks, by comparison, are much more fluid arrangements in which actors may change depending on the issue being addressed. Hugh Heclo described issue networks as being comprised of a large number of participants with varying degrees of commitment and dependence on others. The result, says Heclo, is that networks are sometimes indistinguishable from their environment.[54] One of the key purposes of this work is to evaluate the theoretical arguments about networks, to determine what type of network relationships (if any) were present in Oklahoma City, and to examine the impact those may have had on the response.

The final theoretical argument focuses on the role of culture within and across bureaucracies. James Q. Wilson defined organizational culture as "a persistent, patterned way of thinking about the central tasks of and human relationships within an organization . . . It changes slowly, if at all."[55] Each emergency response domain can be characterized by a socially constructed culture around which organizations and individuals pattern their actions.[56] Culture may be seen as an important factor in mobilizing and motivating bureaucrats in support of their agency's mission, which may contribute to the success or failure of bureaucratic actions. "Organizational culture is typically seen as the glue that binds individuals to an organization's purpose, making the organization, in turn, more efficient."[57] For the purposes of this book, the evaluation will examine the extent to which a unique culture can be observed in the different bureaucracies that participated in the response. The book will then determine what, if any, impact those networks may have had upon the interactions of bureaucracies, across response disciplines and levels of government, as they attempted to respond.

The book will also examine how training may help to develop and reinforce the operational cultures of response bureaucracies. According to Edgar Schein, organizational cultures form in one of two ways: informally through spontaneous interaction in an unstructured group or formally through education and apprenticeship programs.[58] In the case of emergency response, most bureaucracies engage in formal and ongoing training for their employees.[59] These trained experts are the focus of this study. However, it must be acknowledged that there are many people who participated in the response who have no formal training for such activities.

Methods

As with many other bureaucratic activities, evaluating the implementation of response policies is a very difficult endeavor. Emergency response to a catastrophic event is a complicated activity involving the coordinated efforts of a variety of different response bureaucracies (police, fire, health care and emergency medicine, public utilities, government actors, the private sector, volunteers, etc.) across the various levels of government. Each agency may have a different perspective on how the response should proceed and what priorities should be set. The emergency situation is further complicated by the fact that each locality has a unique level of resources and preparations for response activities. Preparation levels vary widely across communities and may have a significant impact on the outcome of the response. The characteristics of each disaster are also unique. The relative consistency between the disasters for which a community is prepared and the one that they actually face may have a significant impact on the perceived success of the response activities.

We do not at this time know specifically what constitutes success in an emergency response to a disaster of great magnitude. Part of this difficulty stems from the fact that we have so few cases from which to learn and draw inferences. Very limited objective material is available that would allow for a quantitative analysis that could aid in differentiating between successful and unsuccessful practices.

More cases are available if one examines natural disasters and accidents. Crisis research has tended to focus on "natural agents, foreign enemies, and sudden disasters" rather than on the unique challenges of a terrorist attack.[60] For some aspects of response, there will be a commonality between the natural disaster and the terrorist incident. An example would be communications. Regardless of whether a major incident is from a natural disaster, an industrial accident, or a terrorist strike, communications can be anticipated to be disrupted in the early phases of the response.[61]

Bureaucratic organizations rely on administrative feedback and program evaluations to guide the formulation of future policies.[62] These tools fundamentally rely on the ability to assess the relative success of the existing policies. However, according to Michael Lipsky, measuring performance within bureaucracies is extremely difficult. He has argued that goal ambiguity, a large number of variables to be considered, and work norms that minimize supervisory scrutiny complicate attempts to measure performance.[63]

Theorist Charles T. Goodsell provided a concise explanation of another reason why it may be difficult to find accurate evaluations of bureaucratic performance.

> For a fundamental feature of bureaucracy is that it continually performs millions of tiny acts of individual service, such as approving applications, delivering the mail, and answering complaints. Because this ongoing mass of routine achievement is not in itself noteworthy or even capable of intellectual grasp, it operates silently, almost out of sight. The occasional breakdowns, the unusual scandals, the individual instances where a true injustice is done, are what come to our attention and color our overall judgment.[64]

This commentary provides important insights into the difficulty of evaluating the success of emergency response. Reports about the response effort may tend to skew any analysis as they may focus on the few negative or ineffective aspects of the response while the many successes may receive little attention.

For this and other reasons, a definitive definition of what constitutes a "successful" emergency response implementation is elusive and not found in either the bureaucracy or the response literatures. In part, this difficulty derives from the general difficulty of defining successful implementation. "Successful implementation can mean many different things to many different people. When policy actors, especially bureaucrats, speak of success they may really have some form of efficiency in mind."[65] However, in the case of emergency response, defining success is much more complicated than identifying whether the actions were efficient. The large number of bureaucracies that are likely to be involved in responding to an act of terror and the variety of tasks that must be carried out by those bureaucracies makes defining success exceptionally difficult. Each response agency also has unique priorities (determined by goals, tasks, and culture), which may complicate coordination of their efforts. Furthermore, the political ramifications of defining a response action as "unsuccessful," when responders themselves may have given their lives in the effort, make it an unpopular option. However, as stated earlier, for the purposes of this study a successful response has been defined as one that limits further damage, casualties, and the impact of the incident on the surrounding communities.

With the variety of agencies that participate, the major challenges they confront, differing opinions about the success or failure of the response across individuals and the difficulty in measuring success, an objective analysis of major emergency response efforts may be elusive. Part of the problem is that members of each response discipline have a unique perspective on the activities at the scene. This may be attributed to the fact that each discipline develops a unique organizational culture that affects their perspective on the events.

However, response to terrorist incidents represents unique challenges to emergency response organizations as the intensity and scope of the incident are not part of the standard operating procedures or routines of the agencies. In fact, the greater the intensity and scope of impact, the less routine the response is likely to be.[66]

Under the conditions of a naturally occurring incident, the local fire department would likely take command of the incident response. However, in a terrorist incident, the command would have to be shared between fire and police to reflect the dual nature of the scene. Furthermore, while emergency response is necessarily a local issue, in a terrorist incident a federal crime has been committed, creating the necessity that federal, state, and local responders coordinate their efforts much earlier in the response.[67] In a natural disaster, there is a slower and more hierarchical introduction of state and federal response elements. According to the Stafford Act, the requirements of the disaster must outstrip the local capacity before local officials can request state assistance. The state then must be able to argue that the response exceeds their capacity before they can request the declaration of a national emergency, thereby mobilizing federal-level response elements.[68] This process can take days. After a major terrorist strike such as the Oklahoma City bombing, where a national emergency was declared by 4:00 p.m. the same day as the incident, the time frame is compressed.

The goal of this study was to utilize expert experiences to test theoretical arguments about bureaucratic implementation of policies. An additional goal was to determine if a structured interview process could yield valuable insights into the relative success and failure of emergency response activities. The expert interviews conducted were the primary source of data for the book. Interviewing has been identified as "[undoubtedly the] most used qualitative method in disaster research" by disaster research scholar Brenda Phillips.[69] Because of the limited number of cases of terrorist incident response, this book takes the approach of examining what the experts who were trained and actually on the scene thought about the emergency response to the Oklahoma City bombing. Those interviewed were not academics or politicians. Instead, the interviews were conducted with expert participants who were at the forefront of the response efforts. Significant insights into the effective and ineffective aspects of the response were gained from operators at the forefront. Their views provided the means to evaluate the theoretical arguments presented.

Several methods were utilized to gather the expert insights. First, an interview questionnaire was developed to create a structured and systematic evaluation. The challenge was to ask questions that would capture the differing views of the experts who had participated in the response from different agencies. For instance, it was anticipated that the views about key response challenges from the fire department, police department, and medical personnel may differ. An additional factor was the desire to develop a questionnaire that would allow for the testing of theoretical arguments about the role of bureaucratic structures, networks, and culture in the response. This was accomplished by the following: incorporating questions to determine the level of government each respondent represented; asking questions to gauge the perceived challenges and successes from each respondent's point of view; and including questions about tasks, plans, and training of each individual. Finally, it was also important that the questionnaire reflect emergency response policies if the success of those polices was to be evaluated in any way. By using the *FEMA Guidelines for State and Local Response Planning*—around which Oklahoma's emergency response plan was patterned—the questionnaire captured the policy element as well as allowing for the creation of a series of discipline-specific questionnaires to get at differing points of view.

The *FEMA Guidelines* organized response into Emergency Support Functions (ESFs), which identify and describe key tasks that may need to be performed in a major incident. State and local governments then assign primary responsibility and supporting responsibility for each task to response agencies. The ESF in the FEMA plan at the time of the incident were direction and control, communications, emergency public information, protective actions, mass care, health and medical, resources management, urban search and rescue, warning, and recovery.[70] Because the Oklahoma City bombing was a terrorist incident for which there was no warning, that category was not included. In addition, because of the focus of this study on the first 12 hours of response, recovery was not utilized as that generally follows after the initial response activities. Each person interviewed was asked a series of general questions about the response and their role in it and then additional questionnaires based on those of the ESF to which their position was relevant.

This book will draw upon the experience and insights of experts who participated in the emergency response to the bombing of the Alfred P. Murrah Federal Building in Oklahoma City, Oklahoma on April 19, 1995. The views of 29 individuals who participated in the response and official reports generated by five different emergency organizations will serve as the foundation for the conclusions drawn by the study. Appendix A provides a list of the people interviewed and their roles in the response. Appendix B provides the reader with the questionnaires utilized in the interview process.

The experts were utilized to guide the researcher, "serving as an insider very well versed in the intricacies of matters within their domain."[71] A consequence of this focus on elite responders is that "passers-by" or "citizen responders" who assisted in the response activities or survivors of the attack were not questioned for their views. This decision was driven by the desire to provide insights into better policy planning for terrorism emergency response. "Civilians" who assisted in the response are not familiar with response policies, what standard operating procedures of response bureaucracies were or how challenges the responders faced were relevant to any policy planning that had been undertaken by city and state officials. Furthermore, they could offer little in the way of informed and practical opinions about the extent to which policies were implemented during the response or what could be done to improve response policies.

Selection of experts for interview was also guided by the desire to focus on the first 12 hours of the response effort. This immediate post-disaster time-period is crucial to achieving a key goal in disaster response: rescue of survivors.

> In Oklahoma they preach that you have to have the ability to respond anywhere in the state within two hours. Within twelve hours, you are not likely to be able to save anyone you have not gotten to. You cannot wait around for the 24–36 hours it may take for the federal government to bring a lot of assets in to assist with the response.[72]

The first 12 hours are a time of uncertainty and critical decision-making—decisions upon which later response activities will be built. This time period is the most difficult and the most essential time following a major disaster both in terms of establishing effective response mechanisms and for rescuing any potential survivors.[73]

Due to the fact that a goal of this work is to advise state and local emergency response planners, individuals who were selected for interview had to have arrived on the scene or engaged in the response (i.e. were not on the site but were mobilizing response elements from their office) within the first 12 hours. Several AARs (discussed below) and one interview were excluded from the analysis because the agency or individual did not arrive on-scene within that first 12 hours.

Another major issue is that many of the key responders, in addition to speaking to other response agencies in the years following the incident, have been interviewed repeatedly in the intervening years. Some of the individuals who were sought for interviews for the current effort declined to participate because of this. Others asked that all of their preceding comments and papers be reviewed before they were interviewed so that they would not be asked the same questions over again. Their wishes were accommodated. This resulted in the need to modify the interview format for some respondents. Finally, events that occurred between the Oklahoma City bombing and the present may have intermingled in the minds of some of the respondents. Specifically, some of those interviewed were called on to participate in the response efforts at the World Trade Center complex after 9/11. When the two responses seemed to get confused it was usually easy to identify the incident to which the respondent was referring based on the unique aspects of each incident. It must be understood that any or all of these factors may have had an impact on the interviews. Ideally, one would conduct this interview process immediately following an incident to overcome these challenges.[74]

One way to help control for these factors was to incorporate a second major type of information for the book. In the year following the incident, there were efforts undertaken to capture the experiences of the response community. First, the City of Oklahoma City utilized a Documentation Team to interview participants for the city's official report on the incident response and for the memorial that was created to commemorate the incident. The memorial made transcripts of some of those interviews available for the book. The goals of that memorial process were different than those of this book. However, the comments of the interviewees were sometimes useful in describing activities or identifying the challenges and accomplishments of the response effort. In some cases the materials provided were not the actual interview but someone's summary of an interview, resulting in the potential loss of the respondent's actual words and ideas. Such documents were treated with skepticism and not integrated with the interview materials.

Another useful form of written source material were the AARs that were filed by most of the major response bureaucracies. An AAR is typically filed for each incident to which a fire department or police department responds. Louise Comfort describes the AAR as "a method of organizational learning that is taken very seriously by the participating personnel. These events provide a way to compare the organization's existing model of performance against the actual requirements of field operations."[75] It provides detailed information about the efforts undertaken to deal with the incident and a timeline of the events that occurred. The AARs were beneficial because they are compiled almost immediately after the incident. For large events such as the Oklahoma City bombing response, multiple individuals from the agencies participate in the effort. This is beneficial in that it allows for multiple perspectives to be included as well as providing some fact checking of each participant's views. Furthermore, the timelines, counts of casualties, equipment received, numbers of individuals participating, etc., are based on written documentation generated on-scene. Such information is generally more reliable than individual recollections, especially after ten years have passed.

Owing to the potential drawbacks of each type of source material, all of the different types of sources were utilized in the research. This approach helped to reinforce the veracity of each source's information. The approach also allowed the researcher to identify potentially inaccurate recollections and disparate points of view. Furthermore, multiple resources were utilized to gain a variety of insights into the response. Clearly, the most significant and valuable contribution was from the interviews with experts at the scene.

A similar research method was used to evaluate human and information infrastructure during the 9/11 response at the World Trade Center. The report was funded by a National Science Foundation grant. It utilized 29 expert interviews, testimony, published reports and articles, news accounts, and television and video documentaries to analyze the role of technology in that response. They found that "effective use of a variety of information technologies helped government agencies better cope with and respond to the multiple crises and ongoing recovery demands resulting from the attack."[76]

An intensive case study analysis was conducted to draw consistent conclusions from the materials and arrive at lessons learned that will help to guide state and local emergency response planners. Those lessons learned will be generated through the identification of patterns and themes in the resource materials. Through this process, the successes achieved and challenges faced by those who responded to this disaster will be discovered. The book will also provide a framework through which other terrorism response activities may be evaluated in the future.

Organization of This Book

This book has been designed to integrate the following: theoretical arguments about bureaucracy; emergency response knowledge; federal, state, and local emergency response plans; and the experiences of emergency responders first on the scene in Oklahoma City. In order to set the context for these arguments, Chapter Two presents a brief summary of the events that unfolded in the first 12 hours after the bombing. The chapter concludes with a table that presents a timeline of the activities during that time period for easy reference.

Once the basic knowledge about the response has been established, the book transitions to the more abstract aspects of understanding response dynamics. Chapter Three delves into the emergency response literature and identifies the challenges that have been identified that are persistently experienced across different response events. It is argued that these anticipated challenges are ones for which the response community should be prepared. That is not to say that these aspects of response should become somehow easy to address. There are some issues that will always be difficult to deal with during the response. However, the responders should have had some foreknowledge of these problems and perhaps have plans for how they might deal with them. The chapter then discusses the activities that were undertaken to address these persistent challenges during the response.

The fourth chapter is more closely tied to the bureaucracy literature. It examines the bureaucratic structures, networks, and culture that had impacts on the response activities. The theoretical concepts are used as analytical lenses through which the response community and activities are examined. The focus of the chapter is on the typical street-level bureaucrat issues and challenges. The expert interviews are used to emphasize and illustrate the arguments and experiences that were present in the theoretical literature and witnessed during the response that day.

Chapter Five takes a third perspective on the response activities. It is more focused on the emergency response plans that were in place at the time of the incident and the way those plans played out in the form of tasks and goals of the participants in the response. The chapter is organized around the federal planning guidelines that were in place at the time of the incident. Specifically, the Emergency Support Functions (ESFs) that identify key response activities and those responsible for them are used to frame the analysis of the way that tasks were executed. Again, the interview materials were used in this chapter to identify and illustrate response activities and challenges.

Finally, Chapter Six provides conclusions in the form of lessons learned and reinforced. A table is provided which identifies the response activities, the agencies identified as responsible for those actions in the state's emergency operations plan, the actions detailed in the plan, and the actions that were taken during the actual response. This helps to highlight consistencies and divergences from the plan, which are then discussed in the concluding text.

Notes

1. Brian Jenkins, "Will Terrorists Go Nuclear?" *Orbis* (Autumn 1985): 511.
2. Amy Smithson and Leslie-Anne Levy, *Ataxia: Chemical and Biological Terrorism Threat and the U.S. Response* (Washington, DC: Stimson Center, 2000), 13–19.
3. John Rielly, *American Public Opinion and U.S. Foreign Policy, 1999* (Chicago, IL: Chicago Council on Foreign Relations, 1999), 7–10.
4. Chicago Council on Foreign Relations, *American Public Opinion and Foreign Policy, 2002* (Chicago, IL: Chicago Council on Foreign Relations, 2003), downloaded from http://www.worldviews.org/detailreports/usreport.pdf on August 4, 2003.
5. See, for example, the first report of the Advisory Panel to Assess Domestic Response Capabilities for Terrorism Involving Weapons of Mass Destruction (also known as the Gilmore Commission) *First Annual Report to the President and Congress: Assessing the Threat* (December 1999); U.S. Commission on National Security in the 21st Century (also known as the Hart-Rudman Commission) *New World Coming: American Security in the 21st Century* (September 15, 1999); GAO, *Combating Terrorism: Opportunities to Improve Domestic Preparedness Program Focus and Efficiency* (Washington, DC: Government Printing Office, November 1998); and K. Jack Riley and Bruce Hoffman, *Domestic Terrorism: A National Assessment of State and Local Preparedness* (a RAND report supported by the National Institute of Justice, U.S. Department of Justice, 1995).
6. See, for example, FEMA, *Consequence Management for Nuclear, Biological and Chemical Terrorism: The Federal Response Plan* (Washington, DC: FEMA, 1996) and the Nunn-Lugar-Domenici Act, Title XIV of the National Defense Authorization Act of 1996 (Public Law 104-201, September 23, 1996), which established training for first responders in selected cities.
7. See, for example, Paul Maniscalco and Hank Christen, *Understanding Terrorism and Managing the Consequences* (Upper Saddle River, NJ: Prentice Hall, 2001), xxx.
8. Louise Comfort, "Risk, Security, and Disaster Management," *Annual Review of Political Science* 8 (2005): 338.
9. Erik Auf der Heide, *Disaster Response: Principles of Preparation and Coordination* (St. Louis: C.V. Mosby Company, 1989), 20.
10. Auf der Heide, *Disaster Response*, 28.
11. David Nice and Ashley Grosse, "Crisis Policy Making: Some Implications for Program Management," in *Handbook of Crisis and Emergency Management*, Ali Farazmand (ed.). (New York: Marcel Dekker, Inc., 2001), 55; and John Kingdon, *Agendas, Alternatives, and Public Policies* (New York: Harper Collins, 1995), 94–100.
12. Thomas Birkland, *Lessons of Disaster: Policy Change after Catastrophic Events* (Washington, DC: Georgetown University Press, 2007), 162.
13. Birkland, *Lessons of Disaster*, 32 and 170.
14. See, for example, Louise Comfort, "Cities at Risk: Hurricane Katrina and the Drowning of New Orleans," *Urban Affairs Review* 41, 4 (March 2006): 503.
15. Thomas Drabek and Gerard Hoetmer, *Emergency Management: Principles and Practice for Local Government* (Washington, DC: International City Management Association, 1991), 30.
16. Brian Toft and Simon Reynolds, *Learning from Disasters* (Oxford: Butterworth-Heinemann, Ltd., 1994), 120.

17. See, for example, Birkland, *Lessons of Disaster*, 159.
18. William Form and Sigmund Nosow, *Community in Disaster* (New York: Harper & Brothers Publishers, 1958), 13.
19. William Waugh, *Terrorism and Emergency Management: Policy and Administration* (New York: Marcel Dekker, Inc., 1990), 76.
20. U.S. Department of Homeland Security (DHS), National Response Framework, January 2008.
21. Ralph Lewis, "Management Issues in Emergency Response," in *Managing Disaster: Strategies and Policy Perspectives*, Louise Comfort (ed.). (Durham, NC: Duke University Press, 1988), 165.
22. Ibid.
23. Comfort, "Cities at Risk," 513.
24. See, for example, Bill Citty, Public Information Officer (PIO), Oklahoma City Police Department (OCPD), interview by author, June 26, 2003; and Ed Hill, OCPD, interview by author, June 24, 2003.
25. The City of Oklahoma City Document Management Team, *Final Report: Alfred P. Murrah Federal Building Bombing April 19, 1995* (Stillwater, OK: Fire Protection Publications, 1996), 365.
26. William Waugh, "Regionalizing Emergency Management: Counties as State and Local Government," *Public Administration Review* 54, 3 (1994): 255; and William Waugh, "Current Policy and Implementation Issues in Disaster Preparedness," in *Managing Disaster: Strategies and Policy Perspectives*, Louise Comfort (ed.). (Durham, NC: Duke University Press, 1988), 111.
27. See, for example, EMSA after-action report, *Terror in the Heartland: The EMSA Response*.
28. See, for example, Debby Hampton, Local Volunteer Manager, American Red Cross, interview by author, June 23, 2003; Joevan Bullard, Assistant City Manager, interview by author, June 27, 2003; and Oklahoma City National Memorial Institute for the Prevention of Terrorism (MIPT), *Oklahoma City, 7 Years Later, Lessons for Other Communities*, 2002.
29. See, for example, Albert Ashwood, Recovery Manager, Oklahoma Department of Civil Emergency Management (ODCEM), interview by author, June 23, 2003; Mike Grimes, Captain, Oklahoma Highway Patrol, interview by author, June 23, 2003; and Gary Marrs, Incident Commander and Chief, Oklahoma City Fire Department (OCFD), interview by author, June 24, 2003.
30. Kenneth Bunch, Assistant Chief, OCFD, interview by author, June 24, 2003; and Bullard, interview by author.
31. See, for example, Gary Davis, Commander, Emergency Medical Service (EMS), OCFD, interview by author, June 27, 2003; Don Stockton, PIO, Oklahoma Highway Patrol, interview by author June 27, 2003; and Jon Hansen, PIO, OCFD, interview by author, June 25, 2003.
32. See, for example, Mike Murphy, Commander, Emergency Medical Services Authority, interview by author, June 26, 2003; Cornelius Young, Major, OCFD, interview by author, June 23, 2003; and Citty, interview by author.
33. See, for example: Spencer Hsu, "After the Storm, Chertoff Vows to Reshape DHS," *The Washington Post*. November 14, 2004: 11; Lara Jakes Jordan, "Former FEMA Director Blames State, Local Officials for Response," *State & Local Wire*. September 27, 2005;

and Norman Ornstein, "Some Steps Congress Can Take to Prevent Another Katrina," *Congress Inside Out*. September 14, 2005.
34. Auf der Heide, *Disaster Response*, 7.
35. William Waugh, "Managing Terrorism as an Environmental Hazard," *Handbook of Crisis and Emergency Management* (New York: Marcel Dekker, Inc., 2001), 670.
36. Robert Agranoff and Michael McGuire, "American Federalism and the Search for Models of Management," *Public Administration Review* 61, 6 (2001): 671–681.
37. Leonard White, *Introduction to the Study of Public Administration* (New York: Macmillan, 1939), 144.
38. Arangoff and McGuire, "American Federalism," 673.
39. Laurence O'Toole, "Multiorganizational Implementation: Comparative Analysis for Wastewater Treatment," in *Strategies for Managing Intergovernmental Policies and Networks*, Robert Gage and Myrna Mandell (eds). (New York: Praeger, 1989), 83.
40. D. Porter, "Federalism, Revenue Sharing, and Local Government," in *Public Policy-Making in the Federal System*, Charles Jones and Robert Thomas (eds). (Thousand Oaks, CA: Sage Publications, 1976), 81–101.
41. Agranoff and McGuire, "American Federalism," 674–675.
42. Walter Kickert, Erik-Hans Klijn, and Joop Koppenjan, "Introduction: A Management Perspective on Policy Networks," in *Managing Complex Networks*, Kickert, Klijn, and Koppenjan (eds). (London, England: Sage Publications, 1997), 1–13.
43. Agranoff and McGuire, "American Federalism," 675.
44. J. Wilson, *Bureaucracy: What Government Agencies Do and Why They Do It* (New York: Basic Books, 1989), 11.
45. Wilson, *Bureaucracy*, 41.
46. B. G. Peters, "The Problem of Bureaucratic Government," *The Journal of Politics* 43, 1 (February 1981): 77–78.
47. U. Rosenthal, C. Michael, and P. Hart, *Coping with Crises: Management of Disasters, Riots, and Terrorism* (Springfield, IL: Charles Thomas Publisher, 1989), 18.
48. J. Pressman, and A. Wildavsky, *Implementation* (Los Angeles: University of California Press, 1984), 164.
49. Wilson, *Bureaucracy*, 12.
50. L. O'Toole, "Multiorganizational Implementation: Comparative Analysis for Wastewater Treatment," in *Strategies for Managing Intergovernmental Policies and Networks*, R. Gage and M. Mandell (eds). (New York: Praeger, 1989), 84.
51. Wilson, *Bureaucracy*, 25.
52. M. Lipsky, *Street-Level Bureaucracy: Dilemmas of the Individual in Public Services* (New York: Russel Sage Foundation, 1980), 3. See also J. Fesler and D. Kettl, *The Politics of the Administrative Process* (Chatham, NJ: Chatham House Publishers Inc., 1996), 288–291.
53. Howard Wiarda, *Introduction to Comparative Politics: Concepts and Processes* (Ft. Worth, TX: Harcourt College Publishers, 2000), 178–179.
54. Hugh Heclo, "Issue Networks and the Executive Establishment," in *The New American Political System*, Anthony King (ed.). (Washington, DC: American Enterprise Institute, 1978), 87–124.
55. Wilson, *Bureaucracy*, 91.
56. Paul Burstein, "Policy Domains: Organization, Culture, and Policy Outcomes," *Annual Review of Sociology* 17 (1991): 328.

57. Mark Cassell, *How Governments Privatize: The Politics of Divestment in the United States and Germany* (Washington, DC: Georgetown University Press, 2002), 107.
58. Edgar Schein, *Organizational Culture and Leadership* (Indianapolis, IN: Jossey-Bass, 2004), 15–21.
59. Maniscalco and Christen, *Understanding Terrorism*, 56.
60. Uriel Rosenthal, Arjen Boin, and Louise Comfort, "The Changing World of Crises and Crisis Management," in *Managing Crises: Threats, Dilemmas, Opportunities*, Rosenthal, Boin, and Comfort (eds). (Springfield, IL: Charles Thomas Publisher, Ltd., 2001), 5.
61. David Wagman, "Emergency Management and Civil Defense," in *Homeland Security: Best Practices for Local Government*, Roger Kemp (ed.). (Washington, DC: ICMA, 2003), 8.
62. Fesler and Kettl, *The Politics of the Administrative Process*, 312–314.
63. Lipsky, *Street-Level Bureaucracy*, 48–53.
64. Charles Goodsell, *The Case for Bureaucracy: A Public Administration Polemic* (Chatham, NJ: Chatham House Publishers, 1985), 37.
65. Randall Ripley and Grace Franklin, *Policy Implementation and Bureaucracy* (Chicago, IL: Dorsey Press, 1982), 87.
66. Gary Kreps, "The Organization of Disaster Response: Some Fundamental Theoretical Issues," in *Disasters: Theory and Research*, E. L. Quarantelli (ed.). Sage Studies in International Sociology (Thousand Oaks, CA: Sage Publications, 1978), 80–81.
67. Maniscalco and Christen, *Understanding Terrorism*, 23–74.
68. T. Robert, Stafford Disaster Relief and Emergency Assistance Act, as amended by Public Law 106-390, October 30, 2000. USC 42, Chapter 68. [As amended by Pub. L. 103-181, Pub. L. 103-337, and Pub. L. 106-390] (Pub. L. 106-390, October 30, 2000, 114 Stat. 1552-1575).
69. Brenda Phillips, "Qualitative Methods and Disaster Research," in *Methods of Disaster Research*, Robert Stallings (ed.). (Philadelphia, PA: Xlibris Corporation, 2002), 205–207.
70. FEMA, *Managing the Emergency Consequences of Terrorist Incidents: Interim Planning Guide for State and Local Governments* (Washington, DC: FEMA, 2002).
71. George Moyser, "Non-Standardized Interviewing in Elite Research," in *Studies in Qualitative Methodology*, Robert Burgess (ed.). (Greenwich, CT: JAI Press, 1988), 114.
72. Bob Ricks, Special Agent in Charge, FBI, interview by author, July 8, 2003.
73. See, for example, Maniscalco and Christen, *Understanding Terrorism*, 42–55.
74. Robert Stallings, "Methodological Issues," in *Handbook of Disaster Research*, Havidan Rodriguez, Enrico Qarantelli, and Russell Dynes (eds). (New York: Springer, 2007), 60.
75. Louise Comfort, "Risk, Security, and Disaster Management," *Annual Review of Political Science* 8 (2005): 344.
76. Sharon Dawes, Anthony Cresswell, and Bruce Cahan, "Learning from Crisis: Lessons in Human and Information Infrastructure from the World Trade Center Response." Unpublished manuscript, 2002.

CHAPTER TWO

Disaster, Chaos, and Response: First Arrival at the Murrah Building Scene

The purpose of this chapter is to establish the basic facts of the emergency response situation encountered in this incident. The chapter provides information about the City of Oklahoma City, its government, the community, and the location of the incident. Once these factors have been established, the chapter turns to a discussion of the bombing and its impact on the surrounding areas. Finally, a brief explanation of the initial phases of the emergency response is presented both as a prose discussion and as a table depicting the timeline of the events that occurred in the first 12 hours of the response. It is intentionally a very brief and basic summary. However, considerable detail about the event and how the response developed is provided in the chapters that follow. This background information is provided in order to provide the reader with the context in which the events under analysis occurred.

Oklahoma City, Oklahoma, is situated in the "Heartland" of America. In the spring of 1995 it was a city of about 450,000 inhabitants. The city is governed by a Council-Manager form of government. This style of governing consists of a Mayor and Council who govern the city, but a City Manager is responsible for the day-to-day functioning of the city.

In 1995, the city's primary emergency response community was comprised of 1,010 firefighters, 995 police officers, and 100 Emergency Medical Services Authority (EMSA) personnel who provided ambulance and emergency medical services (EMS). In addition to these personnel, the city had entered into mutual aid agreements with surrounding cities. Such agreements allowed the city to call upon the emergency response community in surrounding cities in times of need.

The Alfred P. Murrah Federal Building (named after the youngest-appointed federal judge) was situated at Northwest 4th and 5th streets and Harvey and Robinson streets. The building and its parking garage took up a city block. It was designed in 1977 to be highly energy efficient. Its construction was largely concrete, glass, and rebar, with tan brick on the exterior. The building was nine-stories tall with an open floor plan to allow for easy circulation of air. The north face of the building was a glass wall from floors three to nine. The building also featured a series of concrete and steel, load-bearing columns.

The building had a variety of tenants representing functions of the federal government. There were also some state offices and private sector tenants in the building. It housed the following offices: The Department of Agriculture; Department of Defense; U.S. Army recruiters; U.S. Marine Corps recruiters; U.S. Customs Service; Department of Transportation/ Federal Highway Administration; Drug Enforcement Agency; Federal Credit Union; General Services Administration (which actually owned and operated the building); Housing and Urban Development; Secret Service; Social Security Administration; Oklahoma State Water Resources Board; America's Kids Day-Care Center (on the North wall at the windows); and a snack bar. There were approximately 600 employees and about 250 visitors in the building on the morning of April 15, 1995.

The neighborhood surrounding the Murrah Federal Building featured a large number of churches, additional government buildings, and some commercial enterprises. Major facilities surrounding the Murrah Building included the following: Athenian Bar and Grill; Federal Courthouse; First Methodist Church; Journal Record Building; Oklahoma Water Resources Board; Postal Substation; Regency Towers apartment building; Southwestern Bell (numerous sites); St. Anthony Hospital; St. Joseph Cathedral; St. Joseph Rectory; St. Paul's Episcopal; and the YMCA. Figure 2.1 provides a map of the scene.

The explosive was assembled by Timothy McVeigh and Terry Nichols, two Americans who were angry with the federal government and sought to strike out at it.[1] In the legal case that followed the incident, it was determined that McVeigh was the one who detonated the device. While it was demonstrated that Nichols was involved in the making of the bomb, it was established that he was not likely there to emplace the device and arrange for it to explode. This was the primary reason that McVeigh was sentenced to death while Nichols was sentenced to life in prison by both federal and state courts.[2]

An Act of Terror

At 9:02 in the morning on April 15, a 4,800-pound bomb consisting primarily of fertilizer and rocket fuel exploded from a Ryder rental truck parked in the drop-off area in front of the Murrah Federal Building. The blast shattered windows in buildings up to 10 blocks away. It ripped parking meters out of the ground and sent them flying through the air. Cars in the vicinity were smashed by the force of the blast and many of them caught fire, sending plumes of black smoke into the air. The rental vehicle was completely destroyed, with pieces of it found blocks away from the explosion.

The force of the explosion was of such a magnitude that most first responders did not wait to be called out to the incident scene. Rather, they heard and felt it in their offices, cars, and homes and dispatched themselves toward the smoke. Many were going to work or attending morning meetings when the calm was disrupted by the blast in the downtown area. The explosion was felt and heard more than 50 miles away.[3]

The blast devastated the Murrah Building in two major ways. First, it blew in the glass windows and infrastructure on the north face of the building, sending shrapnel flying into the building on all floors. However, a second consequence of the explosion was major structural damage. Several of the supporting columns on the north side of the building were destroyed and others were seriously damaged. The explosion also damaged the crucial steel-support beam that ran across the tops of the columns. Due to the magnitude of the blast, the lower floors of the building were pushed upward. When those floors settled back, the damaged support beam and columns could no longer support their collective weight. The result was a progressive collapse of floors falling downward, accumulating the weight of each successive floor upon those beneath it. In some areas, the floors stacked upon each other in a phenomenon referred to as "pancaking."[4] According to Mike Shannon, who took the lead for much of the search and rescue operations for the Oklahoma City Fire Department (OCFD):

> [in] the pancake area, the average size of a floor was . . . I measured so I would never forget . . . was the distance between the palm of my hand and my elbow. Everything in the floor; all the office furniture, all the copiers, printers, chairs, filing cabinets, desks, everything in that floor would be in that distance including the people.[5]

22 EMERGENCY RESPONSE TO DOMESTIC TERRORISM

Figure 2.1: Layout of the Scene

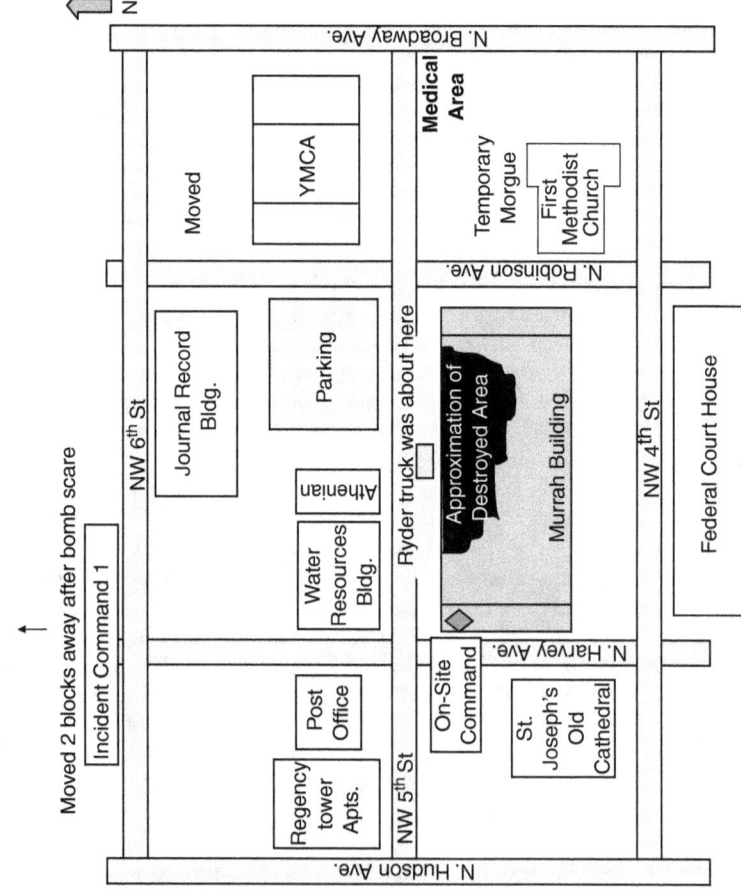

Source: Marrs, Gary. "Report from Fire Chief." *Fire Engineering: Special Issue: Oklahoma City Bombing, Volume 1* (October 1995).

Figure 2.2 provides a graphic representation of the dynamics of the damage done to the Murrah Building. Conspiracy theorists claim that the bomb was not sufficiently large to cause the damage that was done to the building.[6] In fact, the bomb was sufficiently large to undermine key structural supports for the building and the weight of the building itself was responsible for the collapse.

Surrounding buildings also suffered serious damage. Ten buildings, including the Murrah Federal Building, suffered collapse of all or part of their structure. Fifteen buildings had other structural damage. Over three hundred buildings in Oklahoma City had their windows shattered. The City's official AAR also described the area surrounding the building as severely affected by the blast. This included glass and debris littering the surrounding areas, wounded people wandering the streets bleeding and with their clothes in tatters, other people wandering around in shock, and worried people coming into the area searching for their loved ones.[7] Figure 2.3 is a map that was produced by the City of Oklahoma City's Document Management Team. It depicts the geographic spread of the structural damage caused by the explosion on the city as a whole.

The smoke in the area from the burning cars was thick and black. It was so dense that it was difficult for many emergency responders and members of the public to determine exactly what had happened. Firefighters first focused their efforts on the Water Resources Board and the Athenian buildings because the damage to the Murrah Building was not evident from the west end. The heavy smoke also obscured the scene, making damage assessments difficult. Responders arriving from the east headed to the YMCA Building, where it was known that children would be located. As the smoke began to clear, workers saw the damage done by the bomb to the north face of the Murrah Building. Most of the east third of the building was damaged through to the south wall. There was a three-story pile of rubble in the front of the building and a huge crater in the road where the truck with the bomb had been placed.[8]

When first responders arrived on the scene, their first concern was to deal with the wounded and those who were trapped in the damaged buildings and surrounding areas. Between the collapse of the building and the impact of the explosion:

- 163 died in the Murrah Building;
- 1 died in the Athenian Building;
- 2 died in the Oklahoma Water Resources Board building;
- 1 was killed outside, near the blast; and
- 1 nurse was killed when she was hit by falling debris as she tried to help victims.[9]

During the first hours of the response effort, things were chaotic. Many civilians arrived on the scene and entered the buildings in an effort to help. Similarly, emergency response personnel and health care workers converged on the scene. They were under no command structure and many did not have any training on how to rescue victims or deal with the threat of additional collapse of the remaining structure. The civilians proceeded with little concern for their own safety. They had no protective garb or equipment to keep them from becoming victims themselves. Oklahoma City Fire Chief Gary Marrs described the first hour-and-a-half as a major accountability problem because so many people rushed to help out at the scene immediately after the bombing. Volunteers came from widespread areas bearing equipment and supplies to help with the response.[10]

Furthermore, many of the police and other emergency response workers who self-dispatched to the scene were not properly equipped to protect themselves from the dangerous

Figure 2.2: Blast's Impacts on the Murrah Building

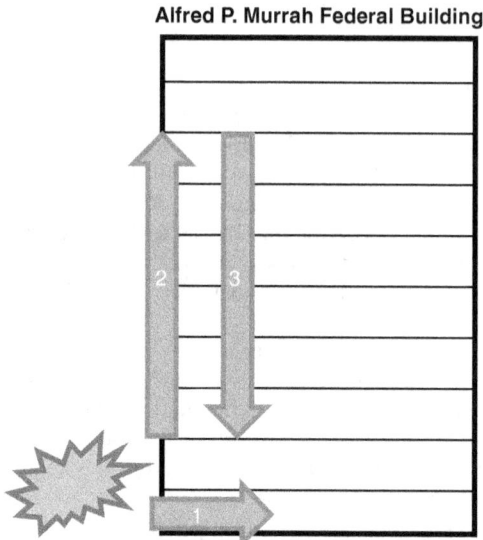

Forces Impact Building

1. Explosive force pushed horizontally through the building. Major damage done to load-bearing columns at the front of the building.
2. Force also pushed floors up and away from the ground
3. When the floors resettled, the support columns were too damaged to hold them. This created a progressive collapse, a situation where the weight of each floor collapsing on the one beneath it results in complete destruction.

Source: Hinmin, Eve. "Explosion and Collapse: Disaster in Milliseconds," *Fire Engineering: Special Issue: Oklahoma City Bombing, Volume 1* (October 1995).

environment inside the damaged buildings. According to the City's AAR, many members of the public, medical workers, and police officers entered the buildings without wearing protective gear of any kind, resulting in some injuries.[11]

However, as the chaos continued, command personnel from all aspects of emergency response were beginning the process of gaining control of the scene. Doing so was a major undertaking. Very rapidly after the explosion there was a major flow of people into the area, including the following: Fire department personnel from Oklahoma City, mutual aid cities, and volunteer firemen; police department personnel from Oklahoma City and surrounding areas; local representatives of the FBI and other federal entities that had offices in the building; Highway Patrol members; Tinker Air Force Base (TAFB) personnel; representatives

Figure 2.3: Extensive Damage Caused to Surrounding Buildings

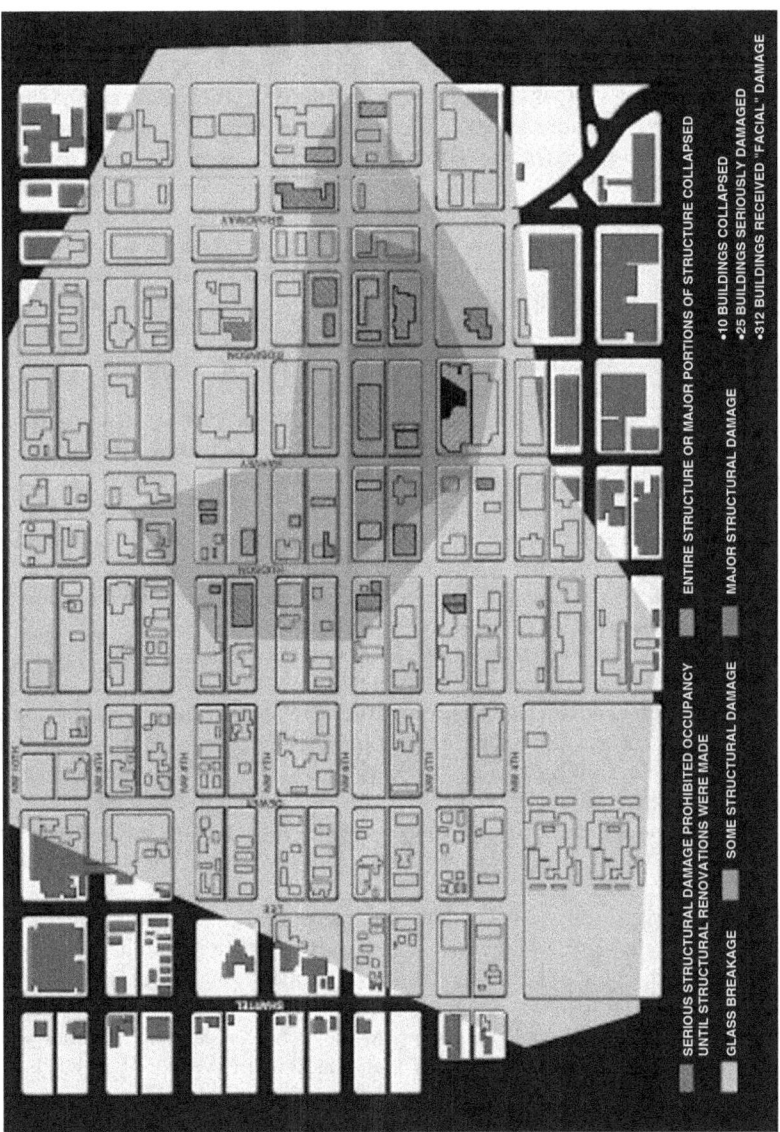

Source: Courtesy of The City of Oklahoma City Document Management Team, *Final Report: Alfred P. Murrah Federal Building Bombing April 19, 1995* (Stillwater, OK: Fire Protection Publications, Oklahoma State University, 1996), 121.

of local media; medical personnel including hospital workers and private practice doctors; EMSA personnel; utility company representatives; the Salvation Army; the Red Cross; private sector entities including crane operators, food providers, phone service providers, and others; and civilian volunteers from the surrounding area.

A huge amount of activity got underway immediately after the explosion. Firefighters, police officers, and many civilians entered the Murrah Building and began to clear rubble in search of any survivors. Of particular concern were the children who had been in the America's Kids Day-Care Center located on the north side of the building at the time of the explosion. Others began the process of searching the surrounding buildings for survivors. Once they had exited or been removed from the building, the wounded sought medical attention from firefighters, EMSA workers, and medical personnel who had arrived at the scene. Triage stations, where the wounded could be prioritized for transport to the hospital based on how badly injured they were, sprung up around the buildings. Many of the wounded got into private vehicles and were transported to hospitals without waiting for the triage process. Firefighters started the work of extinguishing the car fires in hopes of clearing the air so that the responders could get a better idea of what had happened. The police department began the process of establishing a perimeter around the scene so that they could control access to the site and preserve evidence.

In the aftermath of the explosion, the belief was that the incident was caused by the explosion of gas lines in the vicinity. Almost immediately after seeing the crater, trained emergency responders from the fire and police departments determined that the event was caused by an explosion at, rather than under, ground level. This signaled to the response community that this was a criminal act which necessitated treating the area as both a crime scene and a working emergency response.[12] In general, this is a major realization for the responders to make. It changes the nature of the response in that the response command must now be more evenly shared between the Police and Fire personnel.

Chief Gary Marrs of the OCFD, Chief Sam Gonzales of the OCPD, and Special Agent Bob Ricks of the FBI met almost immediately to determine the best way to command the incident. The three friends quickly determined that the fire department would be in charge of the search and rescue operation, the police would be in charge of securing the scene, and the FBI would be in charge of the criminal investigation. This division of labor highlighted the unique nature of the response that was to be undertaken. Owing to the fact that the explosion was an act of terrorism, there were two major dynamics at the scene: one search and rescue operation geared toward finding any survivors and a crime scene investigation focused on finding the perpetrator of the crime.

Members of the private sector mobilized to help with the response. Utility companies arrived on the scene to determine what actions needed to be taken and what support the responders would require. Local construction companies brought cranes and other heavy machinery to the scene to help secure the building and search for survivors. Other local entities brought things like portable bathrooms to the scene, anticipating that they would be required for the long-term response. The Oklahoma Restaurant Association was beginning the second day of its food fair in downtown Oklahoma City at the Myriad Convention Center. They immediately closed down the fair and devoted the building and all the food on hand for the fair to the response effort, serving over 25,000 meals per day for the first 3 days of the response.[13]

Local government members including the Mayor's Office and the City Manager's Office notified the Governor's Office of the incident to begin the process of mobilizing state resources. In turn, the Governor's Office informed the FEMA District 6 office in Texas of the disaster and initiated the process of activating FEMA Urban Search and Rescue (USAR) and Disaster Mortuary (D-MORT) teams to respond to the area. Local government officials also served a function of liaison with the general public for information about the incident. Furthermore, the local officials assisted with the calls needed to bring equipment and supplies to the scene.

Further complicating the response effort was the fact that communications were extremely difficult for the first several hours. Radios, cell phones, and landline phones failed due to the damage caused by the explosion and high call volumes. 9-1-1 services were inundated with calls. "During the first hour of the incident there were over 1,800 calls attempted on 9-1-1 lines alone. At least 1,212 callers received a busy signal due to all incoming trunk lines (15 9-1-1 lines, seven non-emergency lines and two overflow lines) all being used."[14] During the first hours of the response, communications were carried out predominately face-to-face or via runners.

The establishment of order and clearing of unprotected individuals from the scene were facilitated by a startling incident. At 10:30 a.m., there was a bomb scare. A suspicious package was located in the Murrah Building and there was great fear that this was an additional bomb. The order was given to evacuate the building. This posed a significant challenge to first responders and their command structure, as it required responders to abandon people who were trapped in the building. There was conflict between the desire to protect those in the building and to assure the safety of the emergency responders and civilians. However, the evacuation served an important purpose. Once the rescuers and civilians had exited the building, the emergency response commanders were able to gain more control of the scene. They were able to start determining who was in the building, establish more secure perimeters, restrict the civilians from reentering the building, and start to mount a more organized response to the bombing.

After the bomb scare, command and control became more effective on the scene. Activities that began earlier in the morning were continued with a higher level of structure and accountability. Command posts were moved further away from the building in order to distance them from any future threat. Southwestern Bell approached the emergency responders and offered to allow them the use of their facility at their building at Northwest 6th and Harvey Streets. This was deemed an ideal location as it was sufficiently far from the bomb site to be protected, close enough to be convenient, and offered amenities such as bathrooms and a parking deck.

Members of the local media arrived on the scene almost immediately. Once the nature of the incident was established, word got out to state, national and eventually international media outlets. Media personnel continued to converge on the scene throughout the day. Press conferences and interviews began almost immediately and then continued throughout the response. Responders utilized the constant local media coverage as a way to communicate information to citizens. People with loved ones in the area found out which hospitals victims were being taken to and where the children were being transported for pick-up by their parents. Through the media, requests were made for citizens to stay away from the downtown area, for medical personnel to report to hospitals, and for donations of needed goods.

The local medical community mobilized rapidly to triage, transport, and treat the injured. During the first phases of the response, the medical community was challenged by the need to treat a large number of wounded citizens rapidly. There were a total of 675 non-fatal injuries suffered by people in the Murrah Building and surrounding facilities. Hospitals treated lacerations, abrasions, contusions, fractures, dislocations, head injuries, and burns. Eighty-three people required hospitalization for injuries they sustained.[15] The EMSA transported 215 patients to local hospitals, 200 of which were moved within the first 50 minutes.[16] The remainder of the patients were taken to hospitals or doctors' offices in the cars of concerned citizens or walked to nearby hospitals.

FEMA began mobilizing their USAR teams to report to the scene at 10:55 a.m. The FEMA Region 6 Director Buddy Young arrived on the scene by 2:05 to coordinate federal assistance to the disaster. The first FEMA USAR teams began arriving at about 6:30, but the teams were not a major factor in the response until the second day.

At about 4:00, President Bill Clinton signed Emergency Declaration FEMA-3115-EM-OK. Under the Robert T. Stafford Disaster Relief and Emergency Assistance Act, this declaration allowed the federal government to engage fully in the disaster response. It also authorized 100 percent federal reimbursement for allowable expenses incurred during the course of the response. However, while the declaration did free up resources for the City to use as it responded to the disaster, it made only minimal changes to the actual response activities that were occurring on the scene.[17]

Table 2.1 provides a timeline of major response events during the first 12 hours of the response. It is intended to provide an easy reference to the events as they are analyzed through three different lenses (the challenges of response, street-level bureaucratic activities in the response, and the tasks and goals of the response) in the following three chapters.

Table 2.1: Timeline of Events, Oklahoma City, April 19, 1995, 9:02 a.m.–9:00 p.m.[a]

Time	Activity
9:02 a.m.	Bomb exploded at the Alfred P. Murrah Federal Building.
9:03–9:05	Fire stations, police, emergency medical, highway patrol, and Sheriff's Office all self-dispatched to the scene after hearing the blast. Citizen volunteers rushed to the scene to assist in any way they can.
9:05–9:08	OCFD Shift Commander established command at NW 6th and Harvey. Work was undertaken at Athenian Building, Water Resources Board, YMCA, Journal Record Building, and Murrah Building.
9:08	Fire Chief told Dispatch that the Federal Building was the main target and there were car fires to be extinguished.
9:10*	Fire Incident Command established. South triage area established on Harvey east of Murrah Building. Mayor Norick spoke to local media.
9:14	Live television transmission began. All local media provided continuous coverage.
9:15*	First fire companies entered Murrah Building. The ODCEM was fully activated to coordinate the state's response to the disaster and coordinate with FEMA. The EMSA committed 34 units to the response. Mutual response agreements for fire, police, and ambulance companies responded to the scene.
9:19	Police Emergency Response Team activated. Command Post Vehicle ordered to site.
9:20	Additional fire companies brought to the scene.
9:25	Utility companies began to shut off power and assisted in rescue operations.
9:27	First patients were transported from NW 6th and Robinson triage site.

Table 2.1: Continued

Time	Activity
9:30*	Fire department ladders began to be erected at the Murrah Building. The FEMA was notified of disaster. Fire Communications checked all radio towers and transmitter stations. Red Cross, Salvation Army, and Feed the Children responded. Cell phone companies provided phones to emergency workers. Construction companies began providing support.
9:31	Police Command Post Vehicle arrived and was placed at NW 6th and Harvey. The OCPD and others searched for evidence and witnesses.
9:45	First Police Emergency Response Team Squad arrived at Command Post. The OCPD and other law enforcement began to organize perimeter. Oklahoma Restaurant Association began preparing food for rescue workers. Governor Keating made a verbal declaration of disaster.
10:00*	Southwestern Bell Mobile Systems began to work to add channel capacity to downtown cells. FBI established blast-area command post.
10:02	Police E-9-1-1 Center had logged over 1,800 calls.
10:05	Last patients transported from triage—210 had been transported.
10:15*	Oklahoma Highway Patrol and State Bureau of Investigation established command posts. City command structure established. The TAFB Fire Department and others began to arrive on the scene.
10:20	The EMSA began to move NW 6th and Robinson Triage to NW 5th St. The EMSA was told to expect 100–200 more wounded.
10:28	Murrah site evacuated on word of possible explosive in building. Bomb squads evacuate. All downtown offices evacuated.
10:30*	The EMSA established Field Hospital at NE 5th and Oklahoma. The PIOs designated NW 7th and Harvey as Media area. Fire Incident Command evaluated the Murrah Building and organized personnel and equipment for next stage of rescue operation.
10:32	Medical Examiner's staff and OCPD moved first bodies to temporary morgue at the First Methodist Church.
10:45	Police Command Post Vehicle moved to west side of Harvey at NW 9th. Southwestern Bell provided temporary phone lines to the vehicle.
10:55	The FEMA activated Sacramento and Phoenix USAR forces.
11:00*	Fire Command Post relocated to One Bell Central on Harvey. One Bell Central became the main command post. Medical Examiner's Office organized to receive and identify bodies of the dead and collect information on missing persons. Southwestern Bell records show over 12,000,000 phone calls attempted in Oklahoma City.
11:20	The OCPD reported outer perimeter controlled.
1:11	EMSA: Patient transported from the Murrah Building.
1:15	The EMSA closed primary triage unit by Murrah Building; staged ambulances by the building in anticipation of further survivors being found.
1:30	Suspicious crate found in Murrah Building.
1:41	Fire Safety Officer reported he was cleared to begin removing bodies from the building.

(*Continued*)

Table 2.1: Continued

Time	Activity
1:48	Murrah Building evacuated as bomb squads investigate crate.
2:00*	Red Cross conducted initial damage survey and established emergency shelter at St. Luke's Methodist Church. The OCFD EMS established decontamination facilities. The OCFD Logistics organized base in the Journal Record Parking Garage.
2:02	All clear given from bomb scare and rescue operations resumed.
2:05	Region VI FEMA Director arrived in Oklahoma City to coordinate federal assistance.
2:15*	Southwestern Bell Mobile Systems had added channels to its downtown cell sites.
2:30	Rain began.
2:40	Police Command Post vehicle moved to One Bell Central.
2:51	EMSA: Patient transported from the Murrah Building.
2:52	EMSA: Patient transported from the Murrah Building.
2:58	EMSA: Patient transported from the Murrah Building.
3:00*	Generators, light towers, lumber, buckets, and other supplies and equipment arrived at site. Southwestern Bell established site for handling requests for phones in the disaster area.
4:00*	The OCFD and other services organized Critical Incident Stress Defusings for workers. Associated General Contractors of Oklahoma acted as clearing house for donated construction supplies and equipment for contractors working at the site. Police department began issuing entry passes to nonemergency personnel entering the perimeter. City Equipment Services Division went to 24-hour operations to support vehicles used on the site. Refrigerated trucks were provided for use at the temporary morgue and Medical Examiner's area. President Clinton signed an Emergency Declaration 3115-EM officially declaring a federal disaster under the Stafford Act.
5:00*	Funeral Director's Committee opened FAC to receive information from families of missing persons at the First Christian Church. Contractors began to bring heavy equipment to the Murrah Building. Local iron workers provided crews for heavy debris removal operations. Red Cross provided a room and counselors for parents of missing children.
6:00	Oklahoma National Guard contingents began to report for perimeter security. The TAFB increases commitment of personnel, materials, and equipment. AT&T wireless Services COW at Main and Walker was operational.
6:30	Phoenix USAR advance party arrived.
7:00*	Last survivor found trapped in the debris of the Murrah Building. Boldt Construction began first shoring operations. Citizens responded to calls for generators, rain gear, leather gloves, and other items needed for the rescue workers.
8:00	Mental Health volunteers and National Guard Chaplains assisted families at First Christian Church. Workers cleared West Loading Dock for use as Rescue Command Area. Public Works and military units erected tents and shelters for workers.
8:10	FEMA director James Lee Whitt arrived in Oklahoma City.

Table 2.1: Continued

Time	Activity
8:15	Phoenix USAR task force arrived at TAFB.
8:30	Severe thunderstorm in downtown area.

Source: Courtesy of **The City of Oklahoma City** Document Management Team, "Murrah Rescue and Recovery Operation Chronology," in *Final Report: Alfred P. Murrah Federal Building Bombing April 19, 1995* (Stillwater, OK: Fire Protection Publications, Oklahoma State University, 1996), 365–370.

* approximate time.

a This timeline is based largely on **The City of Oklahoma City** Document Management Team's *Final Report: Alfred P. Murrah Federal Building Bombing April 19, 1995*; however, specific alterations were made if there were instances that were not included in the City's timeline. It is a selection of key events in their relatively comprehensive timeline. Other sources used in accumulating this timeline included the author's interviews as well as, Oklahoma City Department of Civil Emergency Management, *After-Action Report: Alfred P. Murrah Federal Building Bombing: Detailed Summary of Daily Activity.*

Notes

1. It was determined that it was important to identify the perpetrators of the incident. However, an exploration of their specific motivations and goals is outside the scope of the current research project.
2. Tony Clark, "Nichols Gets Life for Oklahoma Bombing," *CNN.com*, June 4, 1998, downloaded from http://www.cnn.com/US/9703/okc.trial/nichols.sentence/ on November 20, 2008; and Anonymous, "Nichols Sentenced," *The Daily Oklahoman*, August 9, 2004, downloaded from http://bombing.newsok.com/bombing/history/ on November 20, 2008.
3. City of Oklahoma City Document Management Team, *Final Report*, 9–10.
4. Eve Hinmin, "Explosion and Collapse: Disaster in Milliseconds," *Fire Engineering: Special Issue: Oklahoma City Bombing, Volume 1* (October 1995), n.p.
5. Mike Shannon, Rescue Operations Chief, OCFD, Oklahoma City Documentation Team interview, n.d.
6. See, for example, Steven Yates, "The Oklahoma City Bombing: A Morass of Unanswered Questions," *LewRockwell.com*, May 19, 2001, downloaded from http://www.lewrockwell.com/yates/yates33.html on November 20, 2008.
7. City of Oklahoma City Document Management Team, *Final Report*, 10.
8. Ibid., 11.
9. Injury Prevention Service Oklahoma State Department of Health, *Investigation of Physical Injuries Directly Associated with the Oklahoma City Bombing*, 1996, downloaded from www.health.state.ok.us/program/injury/okcbom.html on May 30, 2003.
10. Gary Marrs, "Report from Fire Chief," *Fire Engineering: Special Issue: Oklahoma City Bombing, Volume 1* (October 1995).
11. City of Oklahoma City Document Management Team, *Final Report*.
12. Ibid., 10.

13. George Buck, *Preparing for Terrorism: An Emergency Services Guide* (Albany, NY: Delmar/Thomson Learning, 2002), 12.
14. OCPD, *After-Action Report: Alfred P. Murrah Federal Building Bombing Incident, April 19, 1995* (1995), 48.
15. City of Oklahoma City Document Management Team, *Final Report*, 339–341.
16. Ibid., 245.
17. ODCEM, *After-Action Report: Alfred P. Murrah Federal Building Bombing, 19 April 1995, Oklahoma City, Oklahoma* (1995), 7.

CHAPTER THREE

Emergency Response Challenges

This chapter establishes the challenges that past emergency response activities would suggest will be encountered persistently during disasters. It draws on the scholarly literature about emergency response as well as governmental assessments of past response activities. This chapter also discusses how the anticipated challenges were specifically encountered in the aftermath of the Oklahoma City bombing. These challenges are things that one might expect response agencies to understand are going to be problematic in the initial phases of a major disaster response. They are aspects of response that, in spite of planning, are likely to remain difficult for the response community to alleviate immediately. For that reason, the fact that an organization experiences difficulties in these areas in the first phases of response should not necessarily be labeled unsuccessful. Rather, it is the nature of the challenges that makes them consistently complicated, in spite of having prepared for them. In many instances, these challenges derive from the bureaucratic structures, networks, and cultures that are the major themes of this book.

This chapter highlights the difficulty of emergency response planning. In spite of having formulated and trained on well-developed response plans, there are problems that will occur. "The unpredictable nature of terrorism underscores the importance of planning even as it makes formulating a perfect attack-response plan impossible."[1] This requires that emergency responders both follow existing plans and engage in improvisation to address any difficulties that may have been unforeseen or which were encountered differently than expected.

Many American cities, however, are not prepared to confront the challenges of responding to acts of terrorism or major natural disasters. Cities lack equipment, sufficient training in WMD response, adequate public health facilities, and training for their emergency responders.[2] Developing a minimum level of local response preparation across the country is in the national interest. However, local preparedness levels vary widely.[3] This is true also in spite of the fact that "government emergency response organizations at all levels prepare contingency plans as part of their operating framework."[4] In many ways, while the plans improve response performance, many challenges persist that could not be anticipated by the plans or that will continue to be difficult to address even if they were anticipated. These were well summarized by William Waugh as follows:

> (1) the diversity of hazards and disasters; (2) low salience of emergency management as an issue; (3) historic resistance to regulation and planning; (4) lack of strong political and administrative constituencies (advocates); (5) uncertain risk from hazards; (6) technical complexity of some regulatory, planning, and response efforts; (7) jurisdictional confusion; (8) economic and political milieu inhospitable to expanding government activities; and (9) questionable capacities of state and local government officials to design, implement, maintain, and operate effective emergency management systems.[5]

These significant impediments to local response planning result in domestic preparedness efforts at the state and local level being largely reactive.[6]

Local response agencies also suffer from the lack of resources common among bureaucracies. One of the ways local governments may prepare to respond more effectively is to form mutual aid agreements (also referred to as interlocal agreements) with neighboring communities. The formation of such agreements is extensive, with most U.S. cities and counties participating.[7] Mutual aid agreements obligate the neighboring community to assist the local responders when the nature of the incident outstrips local capacities. According to the GAO,[8] "these agreements may provide for the state to share services, personnel, supplies, and equipment with counties, towns, municipalities within the state, with neighboring states, or, in the case of states bordering Canada, with jurisdictions in another country."[9] Kurt Thurmaier and Curtis Wood argue that mutual aid agreements reduce uncertainty and allow for financing for personnel, equipment, and services from another jurisdiction in a crisis situation on an as-needed basis.[10] Thurmaier and Wood provide a description of the benefits and motivations to form mutual aid agreements:

> These pacts reduce the level of uncertainty for a jurisdiction more directly by providing the necessary personnel, equipment, or services from another jurisdiction during emergency or crisis situation. Through the use of mutual aid pacts, jurisdictions only need to finance these costs on an as-needed basis.[11]

However, plans and agreements in and of themselves are not a solution to preparedness problems. They work only if those who will be implementing them are familiar with their roles and responsibilities.[12] "The plan itself may be the least important part of the whole disaster planning process. Planning involves meetings and inter-organizational contacts and communications. It involves training exercises and disaster rehearsals."[13] Training and exercises are a major means through which responders can become better able to respond. In fact, Louise Comfort argues that rather than policy guiding practice it is more frequent that policy derives from practice.[14]

Key Anticipated Challenges

Among the difficulties the literature predicts responders will encounter early in a disaster are as follows: the complex nature of emergency incidents; poor communications capabilities; difficulties in coordinating disparate agencies' actions; achieving effective cooperation across the federal layers of government; protecting and effectively utilizing volunteers and goods donated by a mobilized public; managing medical coverage of the response; and addressing the unique characteristics of responding to acts of terrorism. Understanding these anticipated problems is helpful in the effort to evaluate emergency response. A comparison of the problems that were anticipated to those that were actually experienced should reveal whether they were consistent with expectations.

Emergency response is characterized by its complexity, which is marked by the compression of time, the need to act decisively and quickly to keep others from being injured, and the need to coordinate the actions of a variety of actors. Emergency response to a major disaster is seen by some scholars as being the most significant environmental stress that a bureaucracy can face.[15] This sentiment was stated differently by noted response scholar Louise Comfort: "the chief obstacle to improved performance is the enormous complexity of the disaster environment, given the limitations of human cognitive capacity and routine administrative procedures."[16] Crises are also characterized by uncertainty, surprise, and the need for agencies to operate outside of the routine actions for responders.

Chaos

When they arrived on the scene, the Oklahoma City emergency response community had to respond very quickly to a disaster that was at the same time both expansive in its magnitude and concentrated in terms of its geography. In many ways the chaos of the incident scene was the most significant challenge encountered by the response community. This is true in that the chaos made every other challenge more intense, while at the same time each additional challenge encountered made the scene more chaotic.

Mike Shannon, of the OCFD, observed that the first moments of the response were more chaotic and confused than any other response that he had experienced in his career. He stated that the "absolute unrestrained panic[17] was rampant in the building during the first hour to hour-and-a-half of the incident."[18] Other factors, such as the failure of communications or the need to manage the public and media, both contributed to and were derived from the chaotic first day of response. While respondents were not asked whether the scene was chaotic, more than 50 percent spontaneously mentioned the chaos during their interviews. Oklahoma City Fire Chief, Gary Marrs, argued that the chaos was due (among other things) to the lack of control of the scene and the influx of spontaneous rescue activity. He said that the responders knew that it was chaotic and uncontrolled, but he felt there was nothing they could have done to solve the problem immediately.[19] Ed Hill, Captain on the Emergency Response Team for the Oklahoma City Police Department (OCPD), argued that even with a lot of planning for the fact that there will be chaos generated from a major disaster, the best that can be hoped for is an organized chaos where each agency engages in response by doing what they are trained to do.[20]

The chaos that will be a key characteristic of the early hours of a major response contributes to the difficulty in evaluating response.[21] Gary Marrs argued that the emergency response community will never be able to plan or prepare in such a way that they can keep the chaos from happening. However, they can plan for the fact that the chaos will exist and develop means for bringing about order.[22] Ann Burkle reinforced this notion. She said that you have to expect chaos to happen. While emergency rooms all over the country deal with it in a small way on a daily basis, she argued, they still have to plan to deal with the additional chaos that will accompany a major incident.[23] Dan Stockton also argued that chaos will be a factor that cannot be "planned for" but has to be recognized in the planning process. He felt that the chaos is associated with the large numbers of individuals who are compelled to work together in a major disaster where they would not normally. Stockton said that emergency responders are working their way toward means that will allow them to regain order from chaos more quickly.[24]

This environment is produced by the attempts of many different entities to come together to resolve the problem facing society. "When a disaster threatens a community, it requires different responses from different organizations at different locations to set aside prior activities and focus time, effort, and attention on a common goal. To achieve a coordinated response, these actions must be taken simultaneously."[25] In addition to the organizations referred to in this discussion, the coordination must also include individual citizens who come together as a community in the aftermath of disasters.[26]

Overcoming chaos is one of the purposes of the Incident Command System (ICS) (discussed in greater detail in Chapter Four). While chaos reigns on the incident scene, as it will, members of the command staff can be busy on the sidelines preparing their response organization and command structures to regain control. First responders who will be decision-makers in major incidents need to be made aware of the fact that things will be chaotic. They must rise above the chaos and distance themselves from it in order to gain control and orchestrate the response. Thus, those who are making the decisions may have a different perspective of the relative success or failure of the response efforts than those who are "in the trenches." Furthermore, the perception of chaos can change across agencies or at different levels of the bureaucracy or government. Kenneth Bunch related an experience where he was giving a talk on the response one day and was approached by someone else who had been at the Oklahoma City response. He said "chief, I'm wondering if you were at the same incident I was at . . . because from my standpoint it was total chaos." Bunch said that it is chaos, but their charge is to bring some kind of order from that chaos and it is a relative perspective.[27] Dan Stockton of the Oklahoma Highway Patrol agreed. He said that if the responders can regain control of themselves in the first hour and plan for the responsibilities that lie ahead for each agency, they have done well.[28]

While the commanders were already implementing the ICS and were managing a wide variety of tasks, they were still not in control of the chaotic scene and had not established accountability for their responders by the second bomb scare at 10:30. Again, however, that is not to say that they were necessarily effective or ineffective. It is impossible to know if any other responders or response agencies could have gained control any more quickly or more effectively. The second bomb scare was mentioned as helpful by over 59 percent of those interviewed. Fear of a second detonation within the already seriously damaged building, combined with an order to evacuate, drove most volunteers and first responders out of the building. Commanders seized the advantage and locked down the scene. The building and surrounding areas were emptied for about 45 minutes while federal law enforcement officers searched the building. Gary Davis argued that the bomb scare allowed the responders to tighten perimeters, gain control over the incident, and perform other response tasks more effectively. He felt this was one of the best things that happened in the initial phases of the response.[29] Kenneth Bunch said that if the bomb scare would not have happened, they would have had to pull their command staff out of the buildings to have a meeting and organize the response.[30] Gary Marrs also mentioned the benefits of the evacuation, including increasing accountability, moving the command post, creating a staging area, clearing the scene of civilians, and laying out a strategic plan for the next hours of the response.[31]

Absent the bomb scare, it may have been more challenging for the commanders to gain control of the scene. If nothing else, it is likely that the scare expedited the accomplishment of some tasks. Bill Citty felt that they would have eventually set up a secure perimeter without the bomb scare, but it would have taken longer to do so as they would have had to start pushing out people on an individual basis. The bomb scare moved almost everyone out of the area at once.[32]

The discussions of the chaos and bomb scare provide excellent examples of the challenges associated with evaluating disaster response performance. Was the response effective when control was improved after the bomb scare in that commanders were able to take advantage of the opportunity, or were the responders just lucky that it happened? Could chaos have been overcome without the bomb scare? Was the chaos typical or excessive? No definitive answers to these questions exist.

Communications

Communications are crucial to effective emergency response. However, the irony of emergency response communications is that while effective operation is vitally important, it is a "recurring challenge in disaster response."[33] In this section, communications will be addressed in two different ways: communications among the responders, and communications between the responders and the general public.

Among Responders

Communications among responders can be problematic from both technical and organizational/interpersonal perspectives.[34] Technically, the communications means such as radios, cell phones, and landline phones may fail due to damage caused by the disaster itself or because of the enormous volume of communications that occur in the immediate aftermath of a disaster. This kind of technical communications breakdown occurred at the 1993 World Trade Center bombing, the Columbine High School incident, the 9/11 attacks, Hurricane Katrina, and at the Oklahoma City bombing. In each case it has been argued that lives may have been saved if the communications failures had not occurred.[35]

From an organizational or interpersonal perspective, communications among responders during major disasters can be hindered by a variety of factors. Emergency response expert William Waugh described this challenge. He stated that "communicating through the chain of command is made more complex by the additional actors, assumption of new tasks by officials, and the need to assume new, often unanticipated tasks."[36] Organizational and interpersonal communications can also be impacted by differing jargon used by agencies and by cultural differences in perspectives among the responders. These are discussed more fully later in this chapter.

Communications in a major disaster is such a persistent problem for emergency responders that Mike Fagel, an emergency management director said: "where every mission fails is in communication."[37] In fact, there are three major technological problems with communications that are potential hurdles for effective response in the future. First, the disaster contributed to communications problems by damaging vital infrastructure. Among many others who commented on this problem, Ann Burkle, the clinical coordinator of St. Anthony's Hospital Emergency Room which responded to the bombing, stated that phone lines went down and radios did not work. She stated that the biggest source of information for the hospital was the media rather than communications with other response entities.[38]

Second, in the immediate aftermath of the disaster, cellular phone capacity for the area was overwhelmed as people attempted to find out more information about the disaster and their loved ones. Mike Murphy, Commander with the EMSA for the city, claimed that "all the horror stories that you hear about communications basically occurred in '95. It boiled down to runners and face to face [communications]."[39] This sentiment was seconded by Albert Ashwood, the Recovery Manager for the Oklahoma Department of Civil Emergency Management (ODCEM), when he stated that the best communication was to get back in the command post and "stand at the police chief's back pocket and say if you need anything, I am right here" because communicating by other means was "just impossible."[40]

Third, there was an interoperability problem among the radios used by the responders. Interoperability is the capacity for different emergency response agencies (e.g. fire, police, EMS, etc.) at the local level to be able to communicate across agencies. This happens, at least in part, because emergency response agencies tend to purchase their own equipment with their needs and budgets in mind rather than having an overarching scheme that would create interoperability.[41] According to Troy Hale, a member of Plans, Operations and Support for the Oklahoma National Guard, communications were terrible at the incident scene. Hale indicated that the problem was not that people did not wish to communicate, but that there was no equipment that would "cross-talk" across agencies.[42] Ed Hill of the OCPD made a similar argument. He stated that "fire [could not] talk to police . . . police [could not] talk to EMSA . . . and [no one could] talk to the military because the Guard was activated and everybody [was] on different radio bands . . . and so we thought cell phones . . . and cell phones proved to be a problem [because] the cell phone sites were just overloaded."[43]

These communications difficulties caused significant challenges to the emergency response community. In fact, 60.6 percent of those interviewed mentioned communications as being a challenge.

With the Public

Communications with the public were also challenging. The response community needed to deal with an increase in calls from concerned citizens in the immediate aftermath of a disaster. Assistant City Manager Joevan Bullard, Steve Ferreira, and Mayor Ron Norick all stated that their initial functions in the response were focused on fielding the massive influx of phone calls from the public.[44] People wanted information, to give donations, and to find out what they could do to help.

One of the means for communicating with the public was the media. However, while the media can be an invaluable partner for getting information out, managing the droves of media who converged on the disaster scene and providing them with consistent and accurate information were major tasks. This is another problem that is predicted by the emergency response literature.

The media will be drawn to the human tragedy of such an incident. As has been the case in other response activities such as 9/11 and Hurricane Katrina, members of the print and broadcast media will turn out in large numbers and threaten to become part of the problem if they are not managed effectively. More importantly, the members of the response community must become adept at using the media to communicate important information to the public. Erik Auf der Heide argued that the media can be used to assist in the response by conveying instructions, stimulating donations, drawing attention to hazards, minimizing inquiries from loved ones, and providing an alternate source of communications when other systems fail.[45]

Effectively engaging the media was a constant challenge for the first responders. Even the PIOs, who are trained to deal with the media and utilize it as a means to communicate with the public, were amazed at the size of the media presence.[46] There was an almost constant struggle to determine how much information was adequate while trying to cooperate with the media representatives on the scene. Efforts were made in Oklahoma to engage the massive media presence constructively. The goal was to make the response community available to the media on a timely basis so that information could be shared more effectively. According to Jon Hansen, the PIO for the OCFD, the media were called together and informed that the response was going to be active 24×7 and there would be a media officer available to them all the time. Hansen stated that the responders allowed the media personnel to set the times for formal briefings that would accommodate their deadlines and production schedules.[47]

Federalism

Response coordination is further complicated by the need to integrate response agencies at the local, state, and federal level. Federalism is a challenge of the intermingling of disparate bureaucratic structures and cultures. Major jurisdictional challenges have been anticipated to occur. The first argument relevant to federalism is that addressing the emergency is, at least initially, a local responsibility.[48] Regardless of the nature of the incident, the local response community must be prepared to shoulder the responsibilities for emergency action in the initial phases, if not longer. In fact, the emergency response plans and activities of the United States are designed so that the federal government should play a supporting role to the local entities rather than taking the lead or bearing sole responsibility.[49] The anticipated challenge of coordinating response efforts across the federal levels was one of the reasons for the creation of the National Incident Management System (NIMS). According to FEMA, NIMS integrates response practices into a national framework for incident management so that responders at all levels can be effective.[50]

Another major challenge, and a way that the incident was different from a naturally occurring disaster, was that federal emergency response agencies became involved at the immediate outset of the disaster. While this may seem contradictory to the earlier assertion that the federal presence is not felt for the first 12 hours of the response, it is not. The earlier argument is founded on the idea that the federal government will not arrive with its resources and personnel in numbers sufficient to take the pressure off the local-level responders during that time. The current argument addresses the coordination of command immediately after the incident occurred. William (Billy) Penn, the FEMA Public Affairs Officer at the time of the incident, commented that one of the reasons that this incident was challenging and different from a natural disaster was that the timetable for response was expedited considerably. What that meant was that the local-level commanders and federal agents who arrived on the scene in its early stages had to begin to coordinate their response immediately.

Joevan Bullard, Assistant City Manager, stated that the local responders started getting into conflicts with FEMA in the first evening. The reaction to FEMA among local responders was "so you're telling us that you're the pros from Dover and you're here to take over . . . fat chance buddy . . . get out of my way." However, Bullard stated that it was something that they were able to work out.[51] Many additional responders also commented that coordinating response efforts across the levels of federalism was a challenge. According to Kenneth Bunch, Assistant Chief in the OCFD, coordination was difficult because the local response community did not know the role that FEMA would play. He further commented that seven years later, they still felt unsure of the role FEMA should be playing in major incidents.[52]

Developing working relationships and assigning tasks across the levels of federalism was a challenge for the first responders. It seems that the difficulties emerged from a lack of experience with working together and a need to identify what roles each entity was prepared and equipped to play. This is indicative of a lack of networking among the responders at the varying levels of government.

Working Both a Crime Scene and Disaster

The aforementioned challenges are common to all types of emergency response. These will be encountered whether the emergency is a naturally occurring disaster or a man-made tragedy. However, as stated by Sydney Freedberg Jr., "every act of terror is at once a crime, a disaster, and an enemy attack."[53] This characteristic is unique to response to acts of terrorism. It results in a significant overlap of jurisdictions among agencies at each level of government and across levels of government.

> International terrorism brings in whole new dimensions to the decision making, invoking crucial issues of national defense and foreign affairs—dimensions that are largely absent in conditioning the relationships among officials dealing with natural disasters.[54]

Overlapping jurisdictions are more intensely experienced in a terrorist attack because the scene of the incident becomes both a disaster response focused on rescue and recovery efforts as well as a working crime scene where responders seek to protect evidence. This compels differing response agencies to coordinate their efforts more extensively than would be required for a natural disaster.[55]

Uncertainty is frequently pointed out as a significant factor in the difference between a natural disaster and a terrorist incident. Attacks generally come as a surprise, where there is frequently some warning of an impending natural disaster. According to Richard Falkenrath, former Deputy Secretary of Homeland Security and noted terrorism scholar, there are three major areas of uncertainty relevant to a terrorist strike: (1) the extent of the impact of terrorist weapons on civilian populations; (2) a relative lack of warning time; and (3) the inability to predict the psychological impact a terrorist attack will have on the domestic population and people's subsequent reaction.[56]

Finally, Charles Wise and Rania Nader stated that "terrorist attacks impose a new level of social, economic, and fiscal dislocation on the nation and its communities, and they involve the use of many specialized resources that are beyond the capabilities of state and local governments."[57] While it is the case that local response agencies work together regularly for major events, the nature of that collaboration changes in the case of a terrorist attack; more players must be coordinated, different levels of government will come in to play, continued threat of repeated attack may be present, and the capabilities of local government are likely to be outstripped quickly by the response requirements.

The Murrah scene was therefore complex in that it was a crime scene where evidence for convicting Timothy McVeigh and Terry Nichols was located as well as being an ongoing search and rescue effort. While local law enforcement agencies may be utilized to provide security or to set up a secure perimeter at a naturally occurring disaster, it is clear in such instances that the law enforcement personnel are in support of the fire department mission. As mentioned previously, when dealing with the aftermath of a terrorist incident this becomes more complicated. Two sets of agencies were on the scene pursuing two very different, and in some ways contradictory, sets of goals. The process of responding to the emergency necessarily interfered with the protection of evidence. While no law enforcement agency would ever argue that rescuing victims should not be the primary goal of the initial response, they will still experience frustration as evidence is potentially being destroyed.

Albert Ashwood commented on this aspect of the response by saying that there were two separate missions, the investigatory missions of FBI and law enforcement at the same time as you are conducting search and rescue. He stated that this is different from the ways that crime scenes are normally conducted.[58] Other responders, including Mike Grimes and Gary Marrs, stated that FEMA had a difficult time adjusting to the need to maintain the integrity of the crime scene investigation. It is likely that this difficulty resulted from FEMA's lack of experience in working on a scene that was both a disaster and a crime scene.

The agency had yet to develop procedures and train for this new dynamic. This resulted in some lack of coordination in the initial phases of the response as both local and federal responders had to adjust to this dynamic.[59] Balancing the potentially competing requirements of rescuing live victims and protecting evidence was a challenge.

Addressing Volunteerism and Donations

As stated earlier, major disasters tend to bring communities together in support of the response and recovery efforts. Therefore, an additional challenge anticipated in the literature is dealing with a public mobilized to assist in the aftermath of a major incident. Two major issues are deemed likely to be encountered. First, it is expected that the public will turn out to volunteer in large numbers. This includes members of the medical community who will arrive at the scene seeking to be of assistance, police and fire responders who are not on duty but who self-dispatch (referred to as freelancing), and citizens who converge on the scene to offer their help.[60] These volunteers must be managed for their own protection and to minimize the disruption they cause for responders. Some of the volunteers may have skills and experience useful to the response. However, even those individuals who may be of greatest assistance must be identified, accounted for, and protected from dangers on the scene.

Second, the literature predicts that in addition to volunteers, the response community will have to deal with organizing, storing, and utilizing an outpouring of donated goods. These goods will be received from citizens as well as concerned members of the corporate community. Among the observations about excess resources are those by response scholar Russell Dynes who stated that unsolicited supplies:

> (a) normally arrive in volumes far beyond the actual needs; (b) are comprised largely of unneeded and unusable materials; (c) require the services of many people and facilities that could be used for more essential tasks; (d) often cause conflict among relief agencies or among various segments of the population; (e) materially add to the problem of congestion in and near the disaster site; and (f) in some cases may disrupt the local economy.[61]

Donations at Oklahoma City were certainly a challenge for responders to collect, organize, and distribute. Ron Moss said that anything the response community asked for showed up in large numbers. He stated he "probably had 20 guys working under [him] right there that were categorizing different gloves, masks."[62]

The human desire to help those in need created additional challenges for emergency response in Oklahoma City. Both responders and the general public mobilized to try and assist with the response efforts. For members of the first response community, this took the form of responders rushing in to help without being dispatched, creating an accountability problem, and making it difficult to establish an effective chain of command.[63] Furthermore, spontaneous response created difficulties for commanders who needed to figure out what personnel are available at the scene, what their expertise was, and where they were, as well as trying to ensure their safety. Several of the responders interviewed identified this spontaneous response among emergency community as being problematic. They stated that initially the perimeter was porous, resulting in the creation of multiple triage areas, a lack of coordination among responders and a difficulty in organizing the response. However, it was also recognized that to some extent this is human nature and may be unavoidable, and in fact may make positive contributions to the response and save lives. Emergency responders are trained to give aid and they will go where they think they can help and will respond to the calls of those in distress.[64]

In addition to the spontaneous response of emergency personnel, the commanders also had to deal with members of the public who arrived on the scene wanting to volunteer as well as a massive outpouring of donations. While the activities of volunteers and the donations provide much-needed support to the response community, they are a challenge in that they must be managed, protected, and organized by the responders to be safe and effective. If well-organized, they can be of great assistance to responders. If not, they can do more harm than good. Gary Davis argued that at Oklahoma City the equipment and volunteers added to the confusion on the scene because they were not controlled at the perimeter of the response.[65]

Debby Hampton, who organized volunteers for the Red Cross, stated that in addition to people on the scene they were inundated with people who wanted to volunteer or give blood arriving at the Red Cross building and Oklahoma Blood Institute.[66] She stated that "hundreds" of people entered the building to give assistance in the first hours of the response and "thousands" of citizens inundated the Red Cross to volunteer or give blood.[67]

Volunteers appeared on the scene in a variety of different capacities. Some entered the building immediately after the blast and tried to aid those trapped inside. Medical volunteers set up triage stations and provided medical care. First responders self-dispatched to the scene and were technically serving as volunteers as they had not been dispatched to the scene. Some who arrived on the scene got mad when they were not permitted to assist.[68] While those responders who self-dispatched were an organizational problem from a command standpoint, they may have saved lives through their quick action. John Clark commented that no one anticipated the number of volunteers that would show up from other police forces. He said that managing these "volunteers" was an organizational challenge that disaster plans should address.[69]

This challenge is a major irony of emergency response. While it is gratifying and helpful to find that the public is so willing to give aid in times of need, effectively managing that assistance so that it is helpful and not a hindrance (and so that no one gets hurt) is a significant challenge for the responders.

Donations and volunteers are another example of the difficulty in evaluating disaster response activities. While the ability to mobilize resources in support of a major incident response is crucial to success, the problem that tends to be experienced (as was in Oklahoma City) is not a difficulty in acquiring resources but rather in dealing with the volunteers and copious quantity of resources that are spontaneously contributed by individuals and groups effectively.

> The problem of too many resources is coming to be recognized as a pattern which is found, at some time or another, in many disasters. When resources are present in greater amounts than needed, they can greatly complicate the already difficult problems of coordination and communication. In the more extreme cases, excessive influx of resources has even been observed to physically impede activity at the scene.[70]

Furthermore, managing the resources in such a way that they are at the ready when needed, easily located, protected from the elements, and out of the way can be a major undertaking. The influx of donated goods to the Murrah scene was remarkable and a challenge for the responders to track and manage. According to Mike Grimes of the Oklahoma Highway Patrol: "probably our least effective for the first 12 hours was what to do with all the stuff that people were driving up and giving to us. We just were not ready for that . . . to be honest with you."[71]

Similarly, managing the large number of volunteers who arrive on the scene is a resources management problem. Over 66 percent of those interviewed mentioned either volunteers or donations as being a challenge that responders had to address during the initial stages of the response. Mike Grimes commented on the "amazing amount of equipment . . . supplies, and everything that people were bringing" as being both a positive and a negative thing. He said that it was good to have access to the supplies but "it became a nightmare trying to handle it."[72] The official AAR commented that establishing collection sites and inventorying the donated materials were among the major failures of state and local planning.[73] Resources of all kinds arrived in huge quantities. Among many other things, citizens brought batteries, rain gear, food, socks, flashlights, knee and elbow pads, hard hats, gloves of all kinds, medicines, and teddy bears.[74]

One major contribution from the private sector that was recognized as very beneficial was from the Oklahoma Restaurant Association. They facilitated the response and took the burden of feeding and housing responders off the response commanders and the American Red Cross.[75]

The Oklahoma Restaurant Association had just finished their annual conference when the explosion occurred. Subsequently, they quickly established a 24-hour food service operation, at the Myriad Convention Center, to feed all emergency response workers. Eventually, the Myriad was established as a center which met the needs of all personnel responding to the incident. Donated clothing, food, equipment and supplies were available on a 24-hour basis.[76]

Other contributions from the private sector included the following: donation of cellular phones by Southwestern Bell and AT&T; Southwestern Bell donated the use of its facilities; McDonalds, Hardees, Dominoes, and other food providers brought food to the scene; WalMart donated a variety of different supplies to the effort; port-a-potties were donated for use in the response in large numbers; masseuses provided massage therapy for responders at the Myriad Center; and local construction firms provided personnel, supplies, cranes, and other heavy equipment.[77]

Management of both volunteer personnel and donated resources was not an issue that was resolved within the first 12 hours of the incident. The process was ongoing over the first several days before an effective management scheme was in place. It is difficult to assess how well volunteers and donations were handled. While it is a problem that is anticipated by the literature, the emergency response community seems to have been caught off guard. The response community may have anticipated the outpouring of support, but was not ready for it in that magnitude that they experienced.

Establishing a Perimeter

A related challenge that was faced by first responders was the need to establish effective perimeters at the scene. The perimeters serve several functions including protecting evidence, protecting responders from the entrance of someone with evil intent, keeping people who are not trained or appropriately equipped off the scene for their own protection, and helping the responders to have accountability for those who are actively working on the scene.[78] The specific challenges included determining how large the perimeter should be, whether there should be an interior perimeter (closest to the scene to control first responders' access to the building) and exterior perimeter (to keep people relatively far from the scene), and getting the perimeters effectively operating as quickly as possible. Without effective perimeters, citizens, some of whom may wish to do harm to the responders, will continue to flow onto the scene.

According to Sam Gonzales, Chief of the OCPD, the outer perimeter was relatively easy to get set. In contrast, however, the inner perimeter was "extremely difficult" to establish. It took the Oklahoma City first responders 2 hours to gain control of the scene. As mentioned previously, gaining control was facilitated by the second bomb scare. People evacuated the building because they were afraid of additional explosion and collapse. The police department was then able to set up a hard perimeter around the building. This was important to protect civilians on the scene and to gain accountability for those responders in the unstable building. They would not let anyone reenter who did not have on appropriate clothing and have appropriate training for the response effort.[79] According to Mike Grimes, one of the most significant things they learned and accomplished in the first 12 hours of the response was the importance of establishing the inner perimeter which was immediately around the Murrah Building.[80]

Rescuing People from the Building

Extricating victims from the Murrah Building and the damaged buildings surrounding it was also difficult. Initially this task involved helping those who were injured but still able to move on their own to exit the structures and get to a triage station. As the response progressed, the focus shifted to trying to remove victims from the rubble. Bad weather exacerbated the problems associated with these efforts. In addition to the physical challenges encountered, responders also had to deal with the psychological ramifications of the incident and its victims. This included the emotions that drove people to rush into the buildings without regard to their own safety and the fear generated by working in the bombed-out shell of the building.[81]

The dangers encountered and attempts to address the many different problems on the scene all complicated efforts to rescue people from the Murrah Building and the others who were impacted by the explosion.

Mortuary Requirements

Related to the need to rescue living victims is the challenge of recovering, tracking, and identifying the bodies of the dead. According to the Oklahoma Medical Examiner, Ray Blakenely one of the challenges was figuring out how many people were in the building at the time of the bomb—including employees, people with business in one of the offices in the building, and visitors.[82] This was important so that the response community could begin to get counts of the injured, missing, and dead.

Ray Blakenely also stated that one of the tasks the responders dealt with was determining where to put the bodies. Early on in the response, bodies were being pulled out of the building as they were found. Bodies were being laid out on the playground next to the building. Once control of the scene was better established, a temporary morgue was set up in the First Methodist Church. At about the same time it was decided that the bodies of the dead could be left where they were until all living survivors were located.[83]

A final problem associated with dealing with casualties was establishing how the response community should count casualties. Initially there was some confusion because reported counts differed based on who was speaking to the media.[84] Officials were concerned that inconsistent casualty counts would confuse and upset the families of victims as well as the community at large. It would also make it seem as if the responders did not really know what was happening on the scene. Once the problem was identified, a specific policy for the body count was established. While it would seem that this would be an easy thing to do, given the grim nature of the scene, the state of some victims and the many different people working the scene, developing a scheme for counting became very important.

Gary Marrs described the process that was established for the body count. He stated that the responders would not add to the count until the body had been removed from the building. Because of the nature of the incident, as a victim was found a representative of the medical examiner's office was called in to assist in the recovery and help provide for identification and evidence protection. The Medical Examiner had to then determine that it was in fact a new body, because there were body parts that were removed from the building and they wanted to avoid exaggerating the count. The victims were then taken to a temporary morgue east of the Murrah Building and then transferred to the county morgue for positive identification.[85] Each body was accompanied by an FBI agent or other officer from the time it was extracted through the entire process. This helped to assure that no information or evidence was lost from any of the victims of the crime. Marrs said that in support of this strategy he informed the fire department that they were not to do body counts but rather they would rely on the Medical Examiner's office to provide an official count.[86]

Aiding the Victims and Families

Obviously another challenge was attempting to address the medical needs of a large number of victims. This began on the scene of the incident where emergency medical personnel had to triage (sort according to severity of injuries) the injured. Timely triage is crucial to effective response. Ann Burkle and Mike Murphy commented on some of the challenges associated with triage in a major disaster response. Burkle indicated that triage in a major disaster differs from the triage performed on patients on a day-to-day basis. She stated that the assessments have to be done very quickly during a disaster scenario. Patients must be assessed and directed to where they need to go based on less information than one would normally have.[87] Murphy stated that triage is complicated by the fact that there are many people performing triage duties, which may give responders an inaccurate picture of the quantity, nature, and extent of injuries present at the scene. Furthermore, Murphy said that decisions not to treat or transport someone immediately may upset the patient, family members, and friends. This can result in a lack of understanding about the triage process and anger at the responder that one's friend or relation is not being treated more urgently.[88]

The triage process is followed by the need to transport victims, where necessary and possible, to a medical facility for treatment. To perform this activity successfully, ambulances and EMS vehicles need to have clear paths of entrance and egress. These must be established by the emergency responders.[89] Upon arrival at medical facilities, the patients must then be given appropriate care. St. Anthony's Hospital was overwhelmed by the number of patients that arrived. Many of the victims had lacerations from the flying debris caused by the explosion. The clinic area was opened up to help accommodate the influx of patients. This was something that the hospital has practiced in their emergency response exercises.[90]

Finally, there was a challenge in aiding the families of those who may have been in the vicinity of the disaster. Family members converged on the scene, area hospitals, and other relevant public areas seeking word of their kin. In a disaster such as the Oklahoma City bombing, where there were victims in the collapsed building for days after the initial response, responders also had to address the longer-term needs of the families of the victims still trapped in the rubble. This was done in Oklahoma City by establishing a Family Assistance Center (FAC, which is discussed more thoroughly later in the book), which was something that was neither planned for nor practiced by the city before the bombing.[91]

What is impressive about the emergency response to the Oklahoma City bombing was that the response agencies encountered not just one of these challenges but all of them. Through training, positive relationships among the local-level emergency responders and a great deal of innovation and improvisation, the community effectively overcame each of these challenges.

This chapter demonstrated several important factors. First, it showed that many of the difficulties encountered in Oklahoma City's emergency response activities were those that were anticipated by emergency response scholars. These included the chaotic nature of the scene immediately after the incident, poor communications, difficulty coordinating disparate response activities, management of volunteers and donations, and addressing the needs of those impacted by the disaster. This reinforces the literature as well as demonstrating that the challenges themselves were not the result of poor planning or response by the community. Second, the chapter suggests that there were difficulties in addressing some of these problems because of differences in the bureaucratic structures of agencies that participated, by a lack of networking of agencies across the different layers of government, and by differing response cultures within each agency.

Notes

1. James Torr, *Responding to Attack: Firefighters and Police* (San Diego, CA: Lucent Books, 2004), 28.
2. Warren Rudman, Richard Clarke, and Jamie Metzl, *Emergency Responders: Drastically Underfunded, Dangerously Unprepared*. Report of an Independent Task Force Sponsored by the Council on Foreign Relations, 2003, 1–15.
3. Donald Kettl, "Contingent Coordination: Practical and Theoretical Puzzles for Homeland Security," *American Review of Public Administration* 33, 3 (2003): 257.
4. G. Davidson-Smith, "Counterterrorism Contingency Planning and Incident Management," in *Domestic Responses to International Terrorism* (Ardsley-on-Hudson, NY: Transnational Publishers, Inc., 1991), 128.
5. Waugh, "Regionalizing Emergency Management," 255; and Waugh, "Current Policy and Implementation Issues," 111.
6. Arnold Howitt and Robin Pangi, "Intergovernmental Challenges of Combating Terrorism," in *Countering Terrorism: Dimensions of Preparedness*, Arnold Howitt and Robin Pangi (eds). (Cambridge, MA: MIT Press, 2003), 44.

7. Kurt Thurmaier and Curtis Wood, "Interlocal Agreements as Overlapping Social Networks: Picket-Fence Regionalism in Metropolitan Kansas City," *Public Administration Review* 62, 5 (2002): 586.
8. Note that the name of this organization changed to the Government Accountability Office in 2004. At the time of the incident, the name was General Accounting Office and each of the documents referenced for that agency were published under that organizational name. For clarity, the abbreviation is used in the text.
9. GAO, *Combating Terrorism: Intergovernmental Partnership in a National Strategy to Enhance State and Local Preparedness.* March 22. (Washington, DC: U.S. Government Printing Office, 2002), 17.
10. Thurmaier and Wood, "Interlocal Agreements," 586.
11. Ibid.
12. Thomas Drabek and Gerard Hoetmer, *Emergency Management: Principles and Practice for Local Government* (New York: International City Management Association, 1991), 98.
13. Ibid.
14. Louise Comfort, "Risk, Security, and Disaster Management," *Annual Review of Political Science* 8 (2005): 339.
15. See, for example, G. Alexander Ross, "Organizational Innovation in Disaster Settings," in *Disasters: Theory and Research*, E. L. Quarantelli (ed.). Sage Studies in International Sociology (Thousand Oaks, CA: Sage Publications, 1978), 216; and Rosenthal, Boin, and Comfort, "The Changing World of Crises," 10.
16. Louise Comfort, "Designing Policy for Action: The Emergency Management System," in *Managing Disaster: Strategies and Policy Perspectives*, Louise Comfort (ed.). (Durham, NC: Duke University Press, 1988), 4.
17. It should be noted that there is a large literature that argues that panic does not really occur in disaster situations. Definitions of panic frequently include actions that are reckless or selfish by individuals seeking to secure themselves. This has not been the experience at disasters, where citizens frequently risk their own lives to give others aid. In fact, Shannon's description of the "panic" would be inconsistent with most definitions as he is describing the large number of people (both responders and citizens) in the dangerous shell of the Murrah Building who were trying to help others. Similarly, the use of the word chaos in this work could come into question as an inappropriate use of the term due to the judicious and selfless actions of the individuals involved. The term chaos is used, however, as that was the word used repeatedly by the emergency responders interviewed, in the media, and in AARs to describe the immediate aftermath of the explosion. While it lacks academic precision, it accurately reflects the views of the community. For an excellent article on this debate, see Lee Clarke, "Panic: Myth or Reality," *Contexts* 1, 3 (Fall 2002): 21–26.
18. Mike Shannon, "Rescue Operations: Doing Battle with the Building," *Fire Engineering: Special Issue: Oklahoma City Bombing, Volume 1.* October 1995.
19. Gary Marrs, Incident Commander and Chief, OCFD, Oklahoma City Documentation Team interview, October 7, 1999.
20. Hill, interview by author.
21. See, for example, Form and Nosow, *Community in Disaster*, 239, 243–245.
22. Marrs, interview by author.
23. Ann Burkle, Clinical Coordinator, St. Anthony Hospital Emergency, interview by author, June 25, 2003.

24. Dan Stockton, Lieutenant, PIO, Oklahoma Highway Patrol, interview by author, June 27, 2003.
25. Louise Comfort, Yesim Sungu, David Johnson, and Mark Dunn, "Complex Systems in Crisis: Anticipation and Resilience in Dynamic Environments," *Journal of Contingencies and Crisis Management* 9, 3 (September 2001): 145.
26. Lee Clarke, "Panic: Myth or Reality," *Contexts* 1, 3 (Fall 2002): 22.
27. Bunch, interview by author.
28. Stockton, interview by author.
29. Davis, interview by author.
30. Bunch, interview by author.
31. Marrs, "Report from Fire Chief"; and interview by author.
32. Citty, interview by author.
33. Auf der Heide, *Disaster Response*, 79.
34. This is fairly consistent with Louise Comfort's discussion of three sets of communication conditions: technical, organizational, and cultural openness—with the cultural openness not being directly addressed here, but indirectly covered in the discussion of bureaucratic cultures later in the book. Louise Comfort, "Managing Intergovernmental Responses to Terrorism and Other Extreme Events," *Publius: The Journal of Federalism* 32, 4 (Fall 2002): 30.
35. Torr, *Responding to Attack*, p. 34; and Viktor Mayer-Schonberger, "Emergency Communications: The Quest for Interoperability in the United States and Europe," in *Countering Terrorism: Dimensions of Preparedness*, Arnold Howitt and Robyn Pangi (eds) (Cambridge, MA: MIT Press, 2003), 300.
36. William Waugh, "Co-ordination or Control: Organizational Design and the Emergency Management Function," *Disaster Prevention and Management: An International Journal* 2, 4 (1993): 20.
37. Wagman, "Emergency Management and Civil Defense," 8.
38. Burkle, interview by author.
39. Murphy, interview by author.
40. Ashwood, interview by author.
41. Wagman, "Emergency Management and Civil Defense," 8.
42. Troy Hale, Plans, Operations, and Support, Oklahoma National Guard, phone interview by author, June 13, 2005.
43. Hill, interview by author.
44. See interviews: Bullard, interview by author; Steve Ferreira, Non-Commissioned Officer in Charge of Disaster Management, TAFB, phone interview by author, June 21, 2004; and Ron Norick the Oklahoma City Memorial Documentation Team interview, n.d.
45. Auf der Heide, *Disaster Response*, 219.
46. See, for example, Hansen, interview by author; Citty, interview by author; Sue Hale, Assistant Managing Editor, *Daily Oklahoman*, interview by author, June 25, 2003; William Penn, Public Affairs Officer, FEMA, phone interview by author, July 12, 2003; and Stockton, interview by author.
47. Hansen, interview by author.
48. See, for example, Russel Dynes, *Organized Behavior in Disaster* (Lexington, MA: Heath Lexington Books, 1970), 16; and John Carroll, "Emergency Management on a Grand Scale: A Bureaucrat's Analysis," in *Handbook of Crisis and Emergency Management*, Ali Farazmand (ed.). (New York: Marcel Dekker, Inc., 2001), 466.

49. See, for example, Birkland, *Lessons of Disaster*, 183.
50. FEMA. *National Incident Management System*, downloaded from http://www.fema.gov/nims/ on August 3, 2005.
51. Bullard, interview by author.
52. Bunch, interview by author.
53. Sydney Freedberg, "Homeland Defense Effort Breaks Down Walls of Government," *GovExec.com Daily Briefing.* October 19, 2001, 4.
54. Charles Wise and Rania Nader, "Organizing the Federal System for Homeland Security: Problems, Issues, and Dilemmas," *Public Administration Review* 62, S1 (2002).
55. Maniscalco and Christen, *Understanding Terrorism*, 254–261.
56. Richard Falkenrath, "The Problems of Preparedness: Challenges Facing the U.S. Domestic Preparedness Program." BSCIA Discussion Paper 2000-28, ESDP Discussion Paper ESDP-2000-05, John F. Kennedy School of Government, Harvard University, 2000.
57. Wise and Nader, quoting Paul Posner, *Combating Terrorism: Intergovernmental Partnership in a National Strategy to Enhance State and Local Preparedness.* Testimony before the U.S. House, Committee on Government Reform, Subcommittee on Government Efficiency, Financial Management, and Intergovernmental Relations. March 22 (Washington, DC: GAO, 2002).
58. Ashwood, interview by author.
59. Grimes, interview by author; and Marrs, interview by author.
60. See, for example, Form and Nosow, *Community in Disaster*, 239; Auf der Heide, *Disaster Response*, 111–114; and Alan Kirschenbaum, *Chaos Organization and Disaster Management* (New York: Marcel Dekker, Inc., 2004), 39.
61. Russel Dynes, "Interorganizational Relations in Communities under Stress," in *Disasters: Theory and Research*, E. L. Quarantelli (ed.). Sage Studies in International Sociology (Thousand Oaks, CA: Sage Publications, 1978), 56–57.
62. Ron Moss, Rescue Operations Commander, OCFD, interview by author, June 24, 2003.
63. Steven Cohen, William Eimicke, and Jessica Horan, "Catastrophe and the Public Service: A Case Study of the Government Response to the Destruction of the World Trade Center," *Public Administration Review* 62, Special Issue (2002): 30.
64. Murphy, interview by author; Young, interview by author; and Citty, interview by author.
65. Davis, interview by author.
66. Debby Hampton, Local Volunteer Manager, American Red Cross, interview by author, June 23, 2003.
67. Marrs, *Fire Engineering*; Hampton, interview by author; and City of Oklahoma City Document Management Team, *Final Report*, 18.
68. Moss, interview by author.
69. John Clark, Lieutenant, Office of Emergency Management, OCPD, Oklahoma City National MIPT interview, 2002.
70. See, for example, Russel Dynes, "Interorganizational Relations in Communities Under Stress," in *Disasters: Theory and Research*, E. L. Quarantelli (ed.). Sage Studies in International Sociology (Thousand Oaks, CA: Sage Publications, 1978), 56–58; and Auf der Heide, *Disaster Response*, 104–110.
71. Grimes, interview by author.
72. Ibid.

73. ODCEM, *After-Action Report: Alfred P. Murrah Federal Building Bombing: Detailed Summary of Daily Activity*, n.d., 5.
74. City of Oklahoma City Document Management Team, *Final Report*.
75. Hampton, interview with author.
76. ODCEM, *After-Action Report* (n.d.), 6.
77. See City of Oklahoma City Document Management Team, *Final Report*.
78. FEMA, *CONPLAN: United States Government Interagency Domestic Terrorism Concept of Operations Plan* (2001) downloaded from www.fema.gov/pdf/rrr/conplan/cplncvr.pdf.
79. Sam Gonzales, Chief, OCPD, phone interview by author, July 1, 2003.
80. Grimes, interview by author.
81. Citty, interview by author; and Mike Shannon, Rescue Operations Chief, OCFD, Oklahoma City National Memorial Center interview, n.d.
82. Ray Blakenely, Director of Operations, Oklahoma Medical Examiner's Office, Oklahoma City National Memorial Center interview, October 20, 1999.
83. Ray Blakenely, Director of Operations, Oklahoma Medical Examiner's Office, phone interview by author, June 24, 2003.
84. Penn, phone interview by author.
85. City of Oklahoma City Document Management Team, *Final Report*.
86. Marrs, interview by author.
87. Burkle, interview by author.
88. Mike Murphy, Commander, EMSA, Oklahoma City Documentation Team interview, March 21, 2000.
89. Stockton, interview by author.
90. Burkle, interview by author.
91. See author's interviews with Marrs, Blakenely, and Hampton.

CHAPTER FOUR

Response as a Street-Level Phenomenon

This chapter focuses on the challenges that street-level first response bureaucrats face as they attempt to address the implications of a large-scale disaster. It explores the ways in which arguments about bureaucracies can help explain the challenges encountered as emergency response activities take place. The views of the Oklahoma City response community are used to illustrate these concepts. The chapter addresses the major research topics in three ways: evaluation of the function of bureaucratic structures; examination of the importance of networking to successful response; and the presence of bureaucratic culture and its impact on the response.

The first theme of the chapter is that response agency activities at disaster scenes are consistent with theoretical arguments about bottom-up bureaucratic management. This is because of the fact that the street-level response bureaucrats take the lead in determining how the response should unfold—especially in the first 12 hours. As stated by noted emergency management scholars William Waugh and Richard Sylves, "emergency management is a bottom-up process. Capacity building has to begin at the level of first responders who will be responsible for dealing with crises and their consequences until support arrives."[1]

A second focus of the chapter is on the role, type, and formation of networks among emergency response bureaucracies. This section addresses the importance of networking and the building of social capital to effective emergency response. It also discusses the role of training in the formation of bureaucratic culture, which can either help or hinder response activities as each agency develops its own culture which may clash with others' when various agencies have to work together. The role that the ICS may play in overcoming this difficulty is addressed in this chapter. Throughout the chapter, interview materials and research are utilized to support arguments about the response and illustrate the views of the responders who were on the scene that day.

In the course of the interviews, several main themes relevant to the street-level bureaucrat surfaced. Those interrelated themes included the following: the importance of building social capital through networking; the critical role of training and the Oklahoma City Emmitsburg experience; challenges faced by local responders when working with state and federal entities; and maintaining accountability among the street-level bureaucrats. This chapter is devoted to an analysis of these critical, street-level issues.

Whether they are put forth in a formal plan or not, bureaucratic organizations have goals. Such goals can help or hinder implementation based on their relative divergence from the concerns of the implementers.[2] In other words, the more closely response goals resemble the interests and beliefs of the street-level emergency responders (i.e. their bureaucratic cultures), the more effective the goals will be in guiding response.

Goals and tasks may be helpful guides, but frequently the events confronted by bureaucrats compel them to diverge from their established goals. "Because of cognitive limitations and the dynamic quality of our environment . . . there is no way for us to understand at first all the relevant constraints on resources. We can discover and then incorporate them into our plans only as the implementation process unfolds."³ Furthermore, "social scientists are ever sensitive to the fact that much of what occurs in the course of an implementation effort cannot be changed by policy formulators and implementers."⁴ A relevant example of this phenomenon would be the actions undertaken by concerned passers-by who help with response efforts and must be managed by the response bureaucracies but who have no knowledge of response policies, task assignments, or planning.

For these reasons, organizational goals and tasks are sometimes very general in nature to allow the street-level bureaucrat latitude in their dealing with individual cases.⁵ In the case of emergency response, goals must be broad enough to account for the fact that each situation encountered will have specific characteristics that require responders to have broad operational flexibility. The unique requirements facing responders at each disaster scene compel them to have adaptive response techniques. Where policies are more explicit and dictate tasks to the emergency responder, the situation they confront at the incident scene may compel them to alter their approach. This argument coincides with Wilson's expectation that situational imperatives may override goals in defining tasks for the emergency responder.⁶

Within the framework of such goals, bureaucracies begin to develop task assignments for the street-level actors. Tasks can be construed as the bureaucracy's attempt to generate procedures that will help them to reach their goals. These tasks are determined by a number of different factors in addition to the institutional goals. These may include: the circumstances confronting the bureaucrat, the interrelationships among members of the bureaucracy, and prior experiences.⁷ For the purposes of this analysis, it will be assumed that the existing tasks of the bureaucracies have been based upon the culture of the bureaucracies themselves and the City of Oklahoma City Emergency Management Operations Plan that was developed through the use of FEMA's list of operational requirements.⁸

Divergent cultures, goals, and tasks among the local response agencies combine with the challenges of coordinating across levels of government to make emergency response to a major disaster complicated, a fact that is anticipated in the bureaucracy literature.

> Coordination problems become much more difficult in multiorganizational settings, where appeals to hierarchy and authority are likely to be of little use in setting priorities, settling disputes or facilitating interaction. Separate missions, competing legal mandates, distinct constituencies, and competition for resources are obstacles to coordination of the activities of diverse organizations.⁹

Local Response Challenges

Two major categories of local challenges, both anticipated by the first response literature, will be addressed in this section.¹⁰ First, attention will be paid to the difficulties of coordinating the actions of a variety of local-level response organizations with different responsibilities and missions. These issues occurred in spite of the implementation of the ICS (which is discussed later in this chapter). Second, the section will examine the integration of federal-level response entities with the local response elements. Finally, the section will address the human element. At issue is freelancing as well as the large number of spontaneously responding emergency personnel. This tendency for responders to rush to the rescue rather than performing their assigned tasks (i.e. when a police officer enters a building to rescue survivors rather than working to establish a secure perimeter outside the building) will be explored in greater detail.

Local Agency Coordination

As Donald Kettl commented, "at its foundation, homeland security is about one of public administration's oldest puzzles—coordination."[11] Emergency response agencies have varied responsibilities and organizational cultures. Under normal conditions, emergency response organizations are free to follow their unique standard operating procedures, with limited need to coordinate their efforts.[12] When a major incident occurs, the need for coordination poses a significant challenge to each response agency. While it is not the case that these agencies will not have ever worked together to manage day-to-day emergencies, it is the case that the interaction among the agencies is more intense in a major incident. According to Howitt and Pangi, the fact that agencies that do not normally coordinate their activities must do so in a major response results in a lack of formal coordination, unclear lines of authority and a lack of personal ties between agencies.[13] As early as 1958, emergency response scholars had identified the potential challenges of diverse agencies coordinating their efforts. Form and Nosow listed the potential problems as follows:

> (a) the unique structures of the participating organizations differ, (b) the conditions that characterize their community performance normally differ, (c) their perceptions of one another in the community social structure differ, (d) the social characteristics of their members differ, and perhaps most crucial, (e) their cooperative activity cannot become routinized because one cannot simulate disaster adequately, and the tendency is for each organization to preserve its self-identity over its cooperative identity.[14]

Coordinating agency efforts has also been a challenge at the federal level, which, according to May and Williams, "is almost never a simple task."[15]

According to emergency management scholar Russell Dynes, the ability for local response entities to coordinate their efforts for successful response is predicated upon several factors. The first relevant issue is the establishment and maintenance of emergency domains. This deals primarily with the legitimacy each agency has to engage actively in specific response tasks. A second important factor, according to Dynes, is the existence of the network of personal relationships within and across agencies. Similarly, a third issue is the presence and strength of chains of suppliers and clients that can be drawn upon to obtain needed resources as well as the ability to manage those materials. Finally, the overall community organization and relations before the incident is a key factor.[16]

In Oklahoma City, by all accounts, the relationship among the top commanders for each response agency was good. An example of the positive comments about the local-level coordination was given by Bill Citty of the OCPD when asked what he felt was the most effective aspect of the response. He stated that coordination and planning were key factors to the way their response developed. He stated that it is important to get to know the other responders in your community and surrounding areas because "it's much easier to work through conflict and to coordinate those efforts if you know those people."[17] He felt they had done this in Oklahoma City. Ed Hill, also of the OCPD, shared his positive assessment of the local-level relationship. He said that everyone was willing to pitch in to get things done. He was impressed at how well everyone was able to work together.[18]

Figure 4.1: Basic ICS Structures

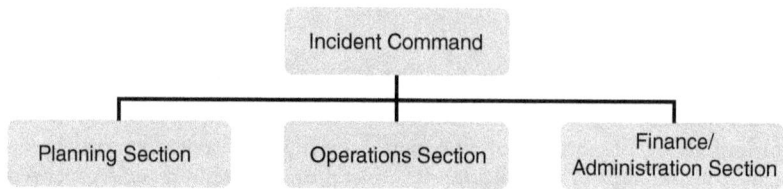

Working under the ICS was also deemed beneficial as it permitted top officials of each responding agency to continue to command their personnel in the accomplishment of their assigned tasks. The ICS is an incident management system that was established in the 1970s to help address the complexities of multiagency response to California wildfires.[19] An organizational chart depicting basic, top tier of ICS structures is presented in Figure 4.1.

Routine operations are generally characterized by minimal resource commitments, brief duration, and not usually being life-or-death in nature. In contrast, incident command in a major disaster is characterized by rapid development of the incident, incomplete information, high stress, and an acceptance that not making a decision is not an option. It is also (as mentioned earlier) characterized by the requirement for multiagency and multidisciplinary coordination. Finally, tasks and coordination requirements will vary based on whether the individual is a first-on-scene responder, a supervisor or a member of the command structure.[20]

Under ICS, each incident, regardless of size, has an Incident Commander. That individual's responsibilities will depend on the size of the incident, as he or she will delegate certain command tasks as the size of the incident expands. It is designed to provide a common organizational structure that is standardized across agencies and scalable to small- or large-scale incidents. Kenneth Bunch emphasized that a benefit of ICS is that "your basic response structure is no different in a 2 story fire over here or a 19 story building that has been blown up."[21]

During small incidents, the Incident Commander will perform all functions. As the incident grows, aspects of the response are factored out to others by the Incident Commander. Other agencies can be added to the response by "plugging-in" to their relevant tasks.[22] This process allows for differing agencies to maintain their managerial autonomy yet still be coordinated with the others in the response. When asked what the most effective aspect of the response operations was, Billy Penn of FEMA argued that it was the Incident Commander, which he indicated provides a central figure to whom all responders can look for coordination and help.[23]

The ICS organizational scheme provides unity of command, common terminology (potentially alleviating an interagency communications problem), and unified resource management. Furthermore, under ICS no one individual is supposed to be responsible for any more than five to seven additional individuals on the scene. This limits the responsibilities of any one supervisor, increases accountability, ensures safety, and helps minimize individuals' tendency to work outside of the command structure (freelance).[24]

The ICS is seen as being flexible enough to be utilized to respond to everyday emergencies, yet robust enough to expand to accommodate the massive influx of agencies and individuals associated with a major disaster.[25] In many ways, what ICS does is help to bring order out of the disorder of early response phases. While the scene may be chaotic, the Incident Commander and others in the command structure can be working on the periphery to develop their organizational scheme. They can then use that organization to impose order on the scene.

The ICS played an important role in the response to the Oklahoma City bombing. Over 46 percent of those surveyed mentioned ICS as a positive influence on the response operations without prompting. A partial depiction of the OCFD's ICS organization for the Murrah incident is included as Figure 4.2. Praise for the role ICS played in the incident came from many agencies including FEMA, the OCFD, the ODCEM, and EMSA. Ron Moss of the OCFD felt that ICS "saved the world." He argued that they use ICS every day and it was helpful in organizing the response that day.[26]

However, not everyone was so positive about the implementation of ICS in general or on that day. Louise Comfort argued that the use of hierarchical tasking (which is an apt description of ICS) to respond to complex problems almost always fails. She stated that hierarchical tasking works reasonably well during routine circumstances but tends to fail under urgent dynamic situations. She contended that such hierarchical systems tend to break down under the cumulative stress of working a major disaster.[27]

At the incident scene, while the system was established and working relatively quickly (with the official timeline establishing the time of initiation at 9:10 a.m.—a mere 8 minutes after the explosion[28]), challenges remained in its operation. In a telephone interview, John Clark stated that in his opinion the least effective aspect of the response was freelancing. He felt this was especially true with the fire department. Clark indicated that this was constructive criticism, but that the fire department will tell you they are using ICS while some fire department members are not. The OCFD had a unified chain-of-command in place at about 10:00 a.m. with assigned responsibilities and duties, but according to Clark the fire department continued to freelance.[29] Mike Shannon said that one problem was that a lot of the people working at the scene that day (including volunteers and other types of response agencies) were not familiar with ICS and this caused misunderstandings.[30] Shannon's view is reinforced by the ODCEM which commented that ICS was weakened early in the response because of the immediate response from numerous agencies, the three locations of the command post within the first few hours, and the deployment of many mobile command posts by a variety of agencies.[31]

In spite of the good will and ICS, management of the response was challenging in part due to the enormity of the event. The OCPD's AAR summed it up from their point of view by stating that the incident was unique because of command being divided among three primary agencies which had to oversee individuals from many additional agencies. For instance, the OCPD AAR stated their commanders had to coordinate the response "of 238 OCPD personnel and 258 officers from 73 municipal agencies, 33 sheriff departments, 8 different state agencies and the National Guard."[32]

Mike Murphy of EMSA stated that communication among agencies was difficult because they operated on different frequencies. But he said that at times the difficulties were not technical. He stated that sometimes the problem was associated with the fact "that police and fire don't necessarily play well together in a command structure in a disaster."[33]

According to Ronny Warren of the OCPD, disputes remain as to what organization should take the lead in responding to a terrorist act. Warren indicated that he feels that the fire department believes that in a terrorist incident they should be in command because of

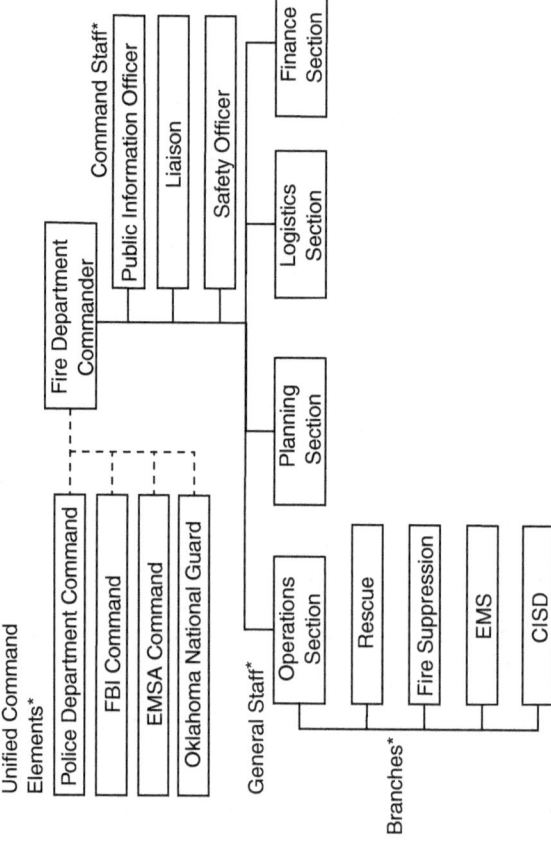

Figure 4.2: Partial Representation of OCFD Murrah Incident Command

*This partial depiction of the Fire Department ICS has been altered from their self-reported organization chart to make it consistent with current ICS.

the hazardous materials threat. However, the police department thinks they should be in charge because it is a crime scene. Warren argued that who should do what and who is in command are big questions. He felt that hazardous materials would be an incident that the fire department would manage. However, he stated that most acts of terrorism are bombings, not hazmat, and guns and bombs, he argued, are the arena of the police department, which would put them in charge. Warren said that these debates are ongoing but feels they are a lot better today at dealing with them than they were years ago.[34]

There were no reports of significant strife among the local responders or comments about local "turf" battles. Overall, it would seem that local coordination was facilitated by existing social networks and the implementation of ICS.

Challenges of Federalism

Response to a major terrorist incident will mobilize responders from the local, state, and federal levels of government. During the interviews, the perspectives about the relations among the local response entities and the federal responders who arrived during the first 12 hours were not as consistent. Gary Marrs, Incident Commander and Oklahoma City Fire Chief, said that in spite of popular belief, he did not "command" the FBI or other law enforcement agencies. Instead, he stated that the responders (fire and police alike) understood their own and the others' responsibilities. They supported each other and kept each other updated on what was happening and what their objectives were for the next time period.[35]

Marrs said that there were some initial problems in integrating FEMA into the response activity, but that these were overcome.[36] Bob Ricks of the FBI also said that they had some control problems with FEMA. He stated that in his opinion FEMA likes to freelance and showed up and tried to take over command. According to Ricks, FEMA initially caused more confusion than assistance.[37] The views voiced by Ricks were reinforced in the interview with Bill Citty.[38]

Other respondents commented that they were stunned to find that the FBI and FEMA, two of the prominent federal-level agencies involved in the response, had some initial difficulties working together. Mike Grimes said that the FBI and FEMA took some time to establish who was in charge, a situation that he stated "almost came into knock-down-drag-out fights." Grimes stated that everyone else at other levels of government knew that the FBI was in charge rather than FEMA because it was a crime scene.[39] Jon Hansen also commented that the various federal agencies (including FBI, Secret Service, Bureau of Alcohol, Tobacco, and Firearms (BATF), Drug Enforcement Administration (DEA), U.S. Marshall's Service, Federal Protective Service) were "very standoffish of each other and didn't blend in very well."[40]

Billy Penn, Public Affairs Officer for FEMA, said that one unique aspect of the response that has been learned from was the need to coordinate FEMA response with the FBI response. Penn said that this was the first time they had had a disaster that was also a crime scene. Normally if there is a crime element, the FBI officer will be in charge. Penn said that after the Oklahoma City response, FEMA took a look at the situation and worked with the FBI to figure out how they should work together if they had both crime scene and disaster again in the future. He said FEMA has spent a lot of time getting to know, working, and planning with the FBI since the incident and that they have worked through the joint organizational response.[41] Based on the comments from interviews and the AAR evaluated for this research, it seems likely that the coordination problem was due to a relative lack of networking at the federal level and a lack of high-level training exercises for major disasters involving a terrorist act.

On the first day of the response, according to Mike Shannon of the OCFD, the problem was rectified when Gary Marrs asserted his position as Incident Commander in the ICS. Shannon said that Marrs told the federal responders that he was the Incident Commander and that this was his community and his building. Marrs told the federal responders that he would be dealing with this incident long after they had gone home so they should accept local command or go home. Shannon said that he felt the emergency response started making good progress after that.[42]

Thus, integrating the efforts of emergency responders across the levels of government posed an early challenge to effective response. However, through the ICS, a federally supported command and control concept, the responders were able to overcome the difficulties and work together relatively effectively.

The Value of Networking

Social capital is a concept in both the public administration and sociology literature that is equally valuable for emergency response. Social capital has been defined by Nan Lin as, the "investment in social relations with expected returns in the marketplace."[43] One acquires social capital by investing time and energy in creating a network of individuals with access to important resources. The network is available to be drawn upon when assistance is required. In other words, the essence of the theory is that "social networks matter."[44] Such networks can be constituted of the interrelationships among individuals or across organizations, which results in two types of capital: human capital and institutional capital, respectively.[45] Therefore, the presence of social capital may allow response personnel and organizations to mobilize additional response capacity in times of urgent need.

According to Nan Lin, social capital of both types can be beneficial in several different ways. Information flows can be generated among members of the network. Influence can be wielded through the social ties among the members of the network. Individuals can gain prestige, recognition, and credentials through belonging to the network. These elements may not be available to the individual absent an investment made in the network.[46]

Social capital, as discussed by Robert Putnam, can be either a private or a public good. It is a private good when individuals utilize it to advance themselves or their goals. However, and more importantly for the purposes of this book, the social capital can be seen as a public good when it is marshaled to bring about a positive outcome for society.[47]

While the term social capital was not used by any of those interviewed, its importance was manifest in the interviews. Half of those interviewed spontaneously mentioned the importance of having built relationships or networked with those within their own organization and, more importantly, across the public and private response entities. Albert Ashwood said that he thought that the key to successful response is "making sure that the people that you are going to be working with on scene are people that you know in the planning process."[48] Social networks were obviously present in Oklahoma City before the incident occurred. In particular, those interviewed mentioned the personal and professional relationship of Fire Chief Marrs, Police Chief Gonzales, and FBI Special Agent in Charge Ricks, who golfed together regularly as well as working together, as being important. This was mentioned by each of those men as well as others who participated in the response.[49] The network among those men helped because there was preestablished trust among the top-level responders as well as general knowledge of what resources each would bring to the response effort. One key way to build such a social network is through training exercises involving all entities that would participate in a major incident response.

Gary Marrs emphasized that the relationship he had with Sam Gonzales and Bob Ricks allowed him to trust them to command their aspects of the response. He had faith that Ricks could command the crime scene and Gonzales could command perimeter aspects of the response competently. In return, Gonzales and Ricks trusted Marrs to command the rescue and recovery. According to Marrs, without that trust, the response would not have worked.[50]

Networking was also deemed an important factor in the relationships between the first responders and private sector entities such as the media, the American Red Cross, and the utility companies. Kenneth Bunch said that he ran into Gary Gardner (whom he had met at a citywide emergency response training in Emmitsburg, MD) from Oklahoma Gas & Electric (OG&E) and it took less than a minute for him to tell Gardner what he needed from him, which could not have happened if he had not have known him. Bunch said that he did not have to explain what the command center was, he just told Gardner that he needed power to the location and Gardner understood its importance immediately. "If you magnify that by 20 times . . . 20 different agencies there . . . it is a benefit to have those relationships built ahead of time."[51] Debby Hampton of the Red Cross and city's final report on the disaster also emphasized the importance of having good working relationships between the private sector and emergency response community in advance of a disaster.[52]

The importance of having a solid working relationship (i.e. social capital or network) among the local media and responders was reinforced by the fact that it was mentioned by all of the respondents relevant to the media including: Sue Hale (Assistant Managing Editor of the *Daily Oklahoman*),[53] Jon Hansen (PIO for the OCFD),[54] and Dan Stockton (PIO for the Oklahoma Highway Patrol).[55]

The value of establishing a network in advance of a disaster was also mentioned by members of the Oklahoma National Guard and the TAFB Fire Department. According to an AAR filed by Adjutant General Stephen Cortright of the Oklahoma National Guard, a "previous joint exercise between [Oklahoma National Guard] and civilian agencies in February 1994 contributed to mission accomplishment."[56] Fire Chief Proctor of the TAFB Fire Department indicated that he had been at training sessions with, and taken classes from, Mike Shannon, who commanded rescue operations for the OCFD. He said that he felt the networking with Shannon ahead of the incident had definitely helped the operation run more smoothly, particularly because Shannon was able to call the Tinker Fire Department immediately after the incident to mobilize their rescue teams and resources.[57]

Also present in Oklahoma was a flexible bureaucratic network among the emergency response entities. During day-to-day operations, the first response agencies carried out their specific roles with limited contact with one another. In this (and other) incidents, however, the various emergency response agencies, members of government, utilities, and private sector entities came together and functioned well together to address the emergency. Evidence of these temporary support arrangements can be found in the presence of mutual aid agreements that were in place between Oklahoma City and surrounding communities. The emergency responders of neighboring communities do not normally participate in Oklahoma City's emergency response functions, but in this case the mutual aid agreements temporarily brought emergency response entities from other cities to assist in the response efforts. This type of flexible and transient network among responders would indicate that issue networks were present rather than the more permanent subgovernment type of management arrangement.

Networking between the local-level officials and federal responders was not as extensive. One means for overcoming the difficulties associated with not having a network with a level of government or agency was suggested in an interview with Kenneth Bunch of the OCFD. He argued that it would be important to establish a presence for the agency not within the network at the command post so that they could be a liaison between the two networks. Having a representative would be helpful because that person will be networked with the people within their own agency that may be able to assist. Without that person, it could take hours of valuable time to identify and locate the appropriate person for each task that needs to be accomplished.[58]

Training

Training exercises are invaluable to response preparation for a variety of different reasons. The relative infrequency with which major responses to terrorist incidents have occurred creates a limited environment for experiential learning. In other words, "disasters are ordinarily rare events, so the [emergency operations plan] is likely never to have been implemented. Exercising is therefore the primary way for the emergency manager to test and evaluate the components of the plan to determine whether they will work in an actual emergency."[59] Training provides the response community with an opportunity to practice implementing their response plans as well as helping to build common norms and values (culture) among the diverse responders. Most cities and states have emergency response plans in place for major disasters. However, those plans are only valuable if all actors in the response are familiar with them and have tested their efficacy.

During training exercises, first responders learn about the response plans and identify potential shortcomings in those plans. Individuals learn what their roles will be in a disaster. According to Ann Burkle, the Emergency Room Coordinator at St. Anthony's Hospital, practicing the plan will help for response activities to become second nature to the first responders.[60] Or, as Cornelius Young of the OCFD recommended, you have got to "train, train, train, train . . . so it just becomes automatic."[61] If the processes "become automatic," this may help to overcome the tendency for peoples' decision-making capacity to decrease in times of stress.[62] Other respondents reinforced the benefits of training. Ed Hill and Kenneth Bunch both argued that creating a written plan, practicing that plan with exercises, and doing advanced trainings on the city's plans are important factors in effective emergency response.[63]

A key benefit of training is that it will allow responders to regain their bearings more quickly and know what they must do when all around them is uncertain. As Dan Stockton said, "it's vastly important that you regain control of yourself before you can become a working member of the team. Otherwise, you're just running around as this individual out here responding."[64]

More importantly, and related to the previous discussion of social capital, first responders learn about each other. In fact E. L. Quarantelli argued that the plan itself may be unimportant. Rather, the meetings and contacts that are made in making the plan and exercising it are important in that they give the opportunity to start networking with other responders.[65] This was expressed by Gary Davis of the OCFD who participated in many training sessions after the bombing. He stated that the social networking associated with training and exercises were actually more important to them than the formal presentations.[66] This sentiment was repeated by Fire Chief Gary Marrs in reference to one of the benefits of participating in a training session at FEMA's Emergency Management Institute at Emmitsburg in July of 1994. He stated that getting the emergency responders for the city together and getting to know one another were key. He said, "I am not so sure that the after-class hours of dinner time and sitting around the campus and just getting to know each other and so forth, wasn't as beneficial as anything was."[67] Police Chief Sam Gonzales was also impressed by the networking opportunity. He said that "the real value [at Emmitsburg] was the relationships that evolved.[68]

In fact, 40 percent of those interviewed mentioned the importance of the Emmitsburg training and its positive impact on the outcome of their response efforts without prompting. The training brought together a wide variety of different response agencies, private sector responders, and members of government to exercise the city's emergency response plan. Joevan Bullard felt that the response was facilitated by the fact that there were about 45 people who went from both the private sector and the public emergency response agencies.[69]

One of the benefits of this type of training is that it took the participants outside of their offices and homes where daily responsibilities could interfere with the training process and placed them in an environment with less interruption. This facilitated networking among the participants, as they were interacting socially before and after the training classes as well as receiving the training itself. Furthermore, the training brought together participants from a variety of different disciplines so that they could become more familiar with each others' roles, responsibilities, resources, and constraints in a major incident response. Ed Hill argued that while most trainings are discipline specific, this one brought together a variety of response agencies in training and scenarios. He thought this formed a stronger bond and taught each participant about what capabilities others had to bring to bear on a disaster.[70]

According to the EMSA AAR, a lack of inclusion of hospitals in the Emmitsburg exercise had a negative impact on their response. The report states that police personnel had to go to hospitals to be certain that hospital radios were on and turned up so that communications could be established with them. Had the hospitals been at the training, according to the report, they would have known to do this.[71] In contrast, Mike Murphy of EMSA stated that a police officer told him that he had learned about the problem of gridlock experienced by ambulances in disasters during the Emmitsburg training. That police officer approached Murphy and asked him which streets he needed opened and guaranteed that those streets would remain clear.[72]

Thus, training provides opportunities for responders to engage in practical learning about their response plans. It gives responders a means through which to test those plans and their ability to implement them. However, responders also benefit from the social networking that accompanies such exercises. An additional benefit of the Emmitsburg training was the opportunity responders had to implement the ICS taught at the Emergency Management Institute for training exercises.

Managing Emotions

Several street-level issues can arise at the early stages of the response that may challenge the command structure as it attempts to gain control. In spite of training and the establishment of the ICS, state and local emergency response planners need to be aware that human emotions will have an impact on the actions of the first responders. In the Oklahoma City incident, there was a rush of first responders into the area and the building. Emergency responders save lives. That is their job. However, it creates the real probability that responders will engage in rescue operations when their role in the incident should have had them performing other tasks. The police department's AAR addressed the issue of emotions rather than plans or procedures influencing responders:

> Given the nature of the disaster, some officers and supervisors left their posts to participate in the rescue operations. Some Officers appeared to become emotionally involved and had difficulty following instructions. It is recommended training continue to emphasize the importance of following directives and maintaining scene security.[73]

Bob Ricks said that there was a desire to participate in the response efforts in everyone in law enforcement, fire, and other emergency response organizations. Ricks stated that he felt that the direction and control people had to let that happen, in spite of the fact that it goes against what they are taught in emergency response textbooks. Each of the participants in the emergency response community had an interest in making sure their own personnel were being taken care of and were safe. Within 12 hours, they had to severely restrict access and only allow specific personnel with certain training access to the building.[74]

Similarly, responders warned that commanders must be careful about the duration of time each individual is engaged in the response activities. While it is important for each member of the response community get an opportunity to participate, commanders must assure that there is a rotation of personnel on and off the scene while maintaining the provision of services for the rest of the city. Furthermore, the passion of the first responders engaged in the incident may cause them to overextend themselves. Incident commanders must assure that responders are sufficiently rested. Mike Grimes commented on the importance of this. He said that it was important that commanders establish a manpower rotation that allows everyone a chance to help in the response but still keeps vital services covered and personnel effective.[75] In a related vein, Ed Hill commented that there was a great deal of determination among those working on the response. He said that workers would be exhausted from working hard but say "give me a cup of coffee and something to eat and point me back towards the building."[76] This emphasizes the importance of the response commanders rotating their work force and instituting and enforcing rotations to assure the safety and efficacy of their responders.

One of the benefits associated with the use of a systematic, scalable command structure (such as the ICS), is that you have greater accountability for the actions of personnel on the scene. Problems of freelancing and overworked personnel can be addressed through this mechanism. Commanders can determine rotation schedules that will allow individuals an opportunity to participate and still provide coverage for other essential services. Furthermore, a small span of control (e.g. number of individuals under the command of another) can help to assure that any individual who is overreaching their capacities or dangerously overworked is given appropriate attention, rest or alternate assignments.

Based largely on interview materials, this chapter evaluated the role of training, bureaucratic culture, and networking on the response activities undertaken in Oklahoma City. Respondents' comments were utilized to emphasize the importance of the ICS, training and exercises, and networks to the success of response to a major disaster. Their responses were also used to address the challenges posed to a major response by the need to activate agencies from each level of the American federal system, and how training, ICS and networking might help to overcome those challenges.

Essentially, the chapter was about the challenges of coordinating the actions of a large number of people who are trained and have acquired cultural beliefs and priorities based on their bureaucracy's mission. The emergency response process can be facilitated by networking, training, and the use of the ICS as each of these helps to integrate the works of different bureaucracies. Where the chapter found difficulties in the response (working with the federal agencies, emotions of the responders, and local agency coordination) these were almost always tied at least in part to insufficient networking or training among the participants. One of the most beneficial aspects of the ICS is that it allows disparate bureaucracies to "plug in" to the response activities and continue to respond as they normally would rather than compelling them to adapt to another agency's way of doing things. Understanding the ways that normal bureaucratic activities are conducted can aid in the creation of effective response plans, help responders to perform their tasks more efficiently, and contribute to the relative success of both day-to-day and major response activities.

Notes

1. William Waugh and Richard Sylves, "Organizing the War on Terrorism," *Public Administration Review* 62, Special Issue (2002): 147.
2. Malcolm Goggin, Ann Bowman, James Lester, and Laurence O'Toole. *Implementation Theory and Practice: Toward a Third Generation* (New York: Harper Collins Publishers, 1988), 127.
3. Pressman and Wildavsky, *Implementation*, 169.
4. Daniel Mazmanian and Paul Sabatier, *Implementation and Public Policy* (Glenview, IL: Scott, Foresman, and Co., 1983), 18.
5. Lipsky, *Street-Level Bureaucracy*, 13–16.
6. Wilson, *Bureaucracy*, 40–43.
7. Ibid., 31–72.
8. City of Oklahoma City, *Emergency Management Operations Plan*, 1994.
9. Edward Jennings and Jo Ann Ewalt, "Interorganizational Coordination, Administrative Consolidation, and Policy Performance," *Public Administration Review* 58, 5 (1998): 418.
10. See, for example, Ali Farazmand (ed.) *Handbook of Crisis and Emergency Management* (New York: Marcel Dekker, Inc., 2001); Auf der Heide, *Disaster Response*; and E. L. Quarantelli (ed.) *Disasters: Theory and Research*. Sage Studies in International Sociology (Thousand Oaks, CA: Sage Publications, 1978).
11. Kettl, "Contingent Coordination," 253.
12. Form and Nosow, *Community in Disaster*, 243.
13. Howitt and Pangi, *Countering Terrorism*, 38–39.
14. Form and Nosow, *Community in Disaster*, 244.
15. Peter May and W. Williams. *Disaster Policy Implementation: Managing Programs under Shared Governance* (New York: Plenum Press, 1986), 38.
16. Russel Dynes, "International Relations in Communities under Stress," in *Disasters: Theory and Research*, E. L. Quarantelli (ed.). Sage Studies in International Sociology (Thousand Oaks, CA: Sage Publications, 1978), 51–63.

17. Citty, interview by author.
18. Hill, interview by author.
19. FEMA, Emergency Management Institute (EMI). *Independent Study Course: ICS Incident Command System* (1998), 1–2.
20. See, for example, Rhona Flin and Kevin Arbuthnot, *Incident Command: Tales from the Hot Seat* (Burlington, VT: Ashgate Publishing Company, 2002), 4 and 38.
21. Bunch, interview by author.
22. FEMA, EMI. *Independent Study Course.*
23. Penn, phone interview by author.
24. FEMA, EMI, *Independent Study Course*, 1–14.
25. Ronald Perry, "Incident Management Systems in Disaster Management," *Disaster Prevention and Management* 12, 5 (2003): 407.
26. Author's interview with Ron Moss.
27. Louise Comfort, "Managing Intergovernmental Responses to Terrorism and Other Extreme Events," *Publius: The Journal of Federalism* 32, 4 (Fall 2002): 33.
28. Oklahoma City Documentation Team, *Final Report* (1996), 366.
29. John Clark, Lieutenant, OCPD, phone interview by author, July 1, 2003.
30. E-mailed questionnaire response from Mike Shannon, Rescue Operations Chief, OCFD, received July 13, 2003. Shannon spoke at length about the response during the interview without actually answering the questions posed. He volunteered to respond to the questionnaire via e-mail.
31. ODCEM, *After-Action Report*, 43.
32. OCPD, *After-Action Report*, 5.
33. Mike Murphy, Commander, EMSA, June, 26, 2004.
34. Ronny Warren, Patrol Sergeant, and Emergency Management Technician with the OCPD, phone interview by author, June 30, 2004.
35. Marrs, "Report from Fire Chief."
36. Marrs, interview by author.
37. Ricks, phone interview by author.
38. Citty, interview by author.
39. Grimes, interview by author.
40. Hansen, interview by author.
41. Billy Penn, Public Affairs Officer, FEMA, phone interview by author, July 12, 2003.
42. Shannon, Oklahoma City Documentation Team interview.
43. Nan Lin, *Social Capital: A Theory of Social Structure and Action* (Cambridge, MA: Cambridge University Press, 2001), 19.
44. Robert Putnam and Kristin Goss, "Introduction," in *Democracies in Flux: The Evolution of Social Capital in Contemporary Society*, Robert Putnam (ed.). (Oxford: Oxford University Press, 2002), 8.
45. Lin, *Social Capital*, 185 and 190.
46. Nan Lin, "Building a Network Theory of Social Capital," in *Social Capital: Theory and Research*, Nan Lin, Karen Cook, and Ronald Burt (eds). (New York: Aldine De Gruyter, 2001), 6–7.
47. Putnam and Goss, "Introduction," 7.
48. Ashwood, interview by author.
49. See interviews with Marrs, interview by author; Gonzales, phone interview by author; and Ricks, phone interview by author. For others see, for example, Bullard, interview

by author; and Ron Norick, Mayor of the City of Oklahoma City, interview by author, June 23, 2003.
50. Marrs, interview by author.
51. Bunch, interview by author.
52. Hampton, interview by author; and Oklahoma Department of Civil Emergency Management, *After-Action Report*.
53. Hale, interview by author.
54. Hansen, interview by author.
55. Stockton, interview by author.
56. Adjutant General Stephen Cortright 1995. Writing in response to an AAR request from the City of Oklahoma City Document Management Team.
57. Fire Chief Proctor, TAFB Fire Department, phone interview by author, July 14, 2004.
58. Bunch, interview by author.
59. Thomas Drabek and Gerard Hoetmer, *Emergency Management: Principles and Practice for Local Government* (Washington, DC: International City Management Association, 1991), 186.
60. Oklahoma City National MIPT, *Oklahoma City*.
61. Young, interview by author.
62. Comfort, "Managing Intergovernmental Responses," 44 citing George Miller, "The Magical Number Seven, Plus or Minus Two: Some Limits on Our Capacity for Processing Information," in *Psychology of Communication* (New York: Basic Books, 1967), 14–44 as well as Rhona Flin, "Decision-Making in Crises: The Piper Alpha Disaster," in *Managing Crises: Threats, Dilemmas, Opportunities*, Rosenthal, Boin, and Comfort (eds), 103–118.
63. Hill, interview by author; and Bunch, interview by author.
64. Stockton, interview by author.
65. E. L. Quarantelli, "The Controversy on the Mental Health Consequences of Disasters," in *Groups and Organizations in War, Disaster and Trauma*, R. Ursano (ed.) as quoted in Drabek and Hoetmer (eds), 98.
66. Davis, interview by author.
67. Marrs, interview by author.
68. Gonzales, phone interview by author.
69. Bullard, interview by author.
70. Hill, interview by author.
71. EMSA, *Terror in the Heartland*.
72. Murphy, Oklahoma City Documentation Team interview.
73. OCPD, *After-Action Report*, 60.
74. Bob Ricks, phone interview by author.
75. Grimes, interview by author.
76. Hill, interview by author.

CHAPTER FIVE

Response Bureaucracies' Tasks and Goals

This chapter begins with a discussion of the role FEMA planning guidelines played in the early phases of response to the Oklahoma City bombing. Tasks and goals can be seen as manifestations of the structures and cultures of response communities.[1] The evaluation of bureaucratic structures and cultures—determining if they were present in the response and, if so, what impact they had—are two of the major research questions of this book. Specific bureaucratic agencies are given responsibilities that are then incorporated into the bureaucratic culture through the training process. Table 5.1 lists the key tasks performed by emergency response agencies at the Oklahoma City bombing in the first 12 hours of the incident response.[2]

The discussion then presents information about the various tasks performed during the response. The major categories of tasks will be formatted (consistent with the questionnaire) based on the following ESFs: direction and control; communications; emergency public information; protective actions; mass care, health, and medical; resources management; and urban search and rescue. In each section, a description of the relevant tasks and goals will be provided as well as respondents' comments about the success of the actions.

Table 5.1: Key Tasks Performed by Participating Response Organizations Based on Interviews and AAR

Organization	Key Tasks Performed (First 12 Hours)*
OCFD	Extinguished fires; search and rescue (Murrah and surrounding buildings); established ICS provided medical assistance to victims; provided public information; incident command for search and rescue; debris removal; donation and volunteer management
OCPD	Search and rescue; evidence protection and collection; perimeter security; established command; medical assistance; provided public information; incident command for perimeter security; debris removal; donation and volunteer management
FBI	Evidence protection and collection; incident command for investigation; liaison with national-level FBI; debris removal
EMSA	Triage patients; transport; provided medical assistance; debris removal; volunteer management
Hospitals	Provided medical care to victims; donation and volunteer management
Mayor's Office	Provision of public information; resource allocation; liaison with Governor's Office; donation management
Governor's Office	Provision of public information; liaison with federal government; declared disaster; resource allocation; mobilized National Guard

Table 5.1: Continued

Organization	Key Tasks Performed (First 12 Hours)*
City Manager's Office	Provision of public information
Oklahoma Highway Patrol	Perimeter security; medical assistance to victims; dignitary protection; debris removal
TAFB and Tinker Fire Department	Provision of resources; transport of FEMA USAR teams; medical assistance; search and rescue; donation management
National Guard	Perimeter security; medical assistance; provision of resources; evidence collection; donation management
Salvation Army	Provision of food
American Red Cross	Provision of food; volunteer registration; sheltering; collecting information about lost children from families; donation and volunteer management
Medical Examiner's Office	Documentation of bodies; analysis of remains; helped establish the FAC; establish temporary morgue; acquire resources
Oklahoma Restaurant Association	Provided food to responders; provided rest stations for responders; donation management

*Author's compilation based largely on interviews.

Task and Goals Based Planning

Emergency response plans put forth the tasks and goals for the various entities that may participate in response efforts. In fact, the National Response Framework (which was the Federal Response Plan at the time of the incident) as well as FEMA guidelines for state and local planning to respond, whether to a terrorist incident or naturally occurring disaster, are not agency-specific. Rather, the plans are focused on tasks that must be performed regardless of the cause of the incident or its magnitude.[3] Albert Ashwood, of the ODCEM, accurately described the FEMA guidelines as not being a detailed plan but, rather, a document that indicates who is in charge and who the support agencies are for each task. He stated that he felt this is all you really need from the plan.[4] This response is consistent with noted emergency response scholar William Waugh's comment that "the model, in other words, is a guide for policy design and implementation rather than a mechanism for action."[5] In many ways, this lack of specificity for actions to be carried out facilitates response in that it allows for improvisation so that the responders can react to the specific characteristics of the disaster.

A comparison of the emergency response plan in place for the City of Oklahoma City at the time of the incident with the *State and Local Guidelines for Emergency Response Planning* demonstrated that local plans were consistent with the guidelines set forth by FEMA.[6] However, when asked if they were familiar with FEMA response guidelines, only 14 respondents (53.8 percent) recalled having known about FEMA response plans at the time. Over three-quarters of those who were familiar with the guidelines felt that the response complied either somewhat or fully with the FEMA guidelines. However, views about the extent to which the response corresponded with plans varied. For instance, Gary Marrs indicated that they had trained with their plan and he felt that the response was fairly consistent with it. He emphasized that the response was not 100 percent in compliance with the emergency response guidelines, but that it "conformed with a majority of the standards somewhat to fully."[7] Albert Ashwood felt that they were a little off the plan but that there was a communications gap that hindered its implementation until they got it worked out.[8] Mike Murphy argued that the ICS was more influential over the city's response than the emergency support functions found in the FEMA Plan.[9] Interestingly, ICS is recommended for use by FEMA to aid in the Command and Control ESF.[10]

At the time of the incident, John Clark of the OCPD felt that the police department's response was not conducted in compliance with the FEMA guidelines. However, after the event he was transferred to the police department's planning division. In that capacity he became more familiar with the FEMA planning guidelines and thought that, looking back on the response effort, the actions undertaken were consistent with FEMA's guidance.[11]

Sam Gonzales felt that the responders were familiar with the FEMA guidelines and what they should be doing, but that they "threw all that out the window" during the response effort. Gonzales said he felt that the response plan did not work in this case because there was so much devastation in a very small area. Furthermore, Gonzales stated that he felt the FEMA guidelines were more appropriate to the fire department responsibilities and missions than that of the police department.[12]

Mike Grimes of the Oklahoma Highway Patrol agreed that the FEMA guidelines were not the basis for the response activities. He was the Federal Emergency Response Liaison, but he argued that they were operating under the Highway Patrol guidelines and federal response planning guidelines were not even a consideration.[13]

Based on the views expressed by those familiar with the emergency plans, it seems that while the plan itself may not have been implemented, there was general agreement that having had a plan that set forth tasks and goals for the responders was valuable. It also seems that a plan needs to be structured in such a way as to allow state and local planners and responders to tailor it to their specific needs and capabilities.

Finally, it is interesting to note that the majority of those interviewed (78.6 percent) felt that the tasks they performed in the first 12 hours of the response were either somewhat related to or corresponded well with the tasks they would expect to perform in a major disaster response. This indicates that the planning and training that occurred before the incident were beneficial to the responders in that it helped them to have realistic expectations of their roles.

Direction and Control

Direction and control is the process by which management of the actions of the responders is coordinated and commanded. These activities were carried out by the local first response community which then integrated state and federal responders as they arrived on the scene. Among the goals of this support function are to gain control of the scene and those working it, to mobilize the resources that will be required for the response, to achieve accountability for the individuals working the scene, and to bring order out of chaos. FEMA recommends that the ICS be utilized to aid in this process.[14]

Effective command is crucial to the success of response efforts. Bob Ricks of the FBI stated that you need to have one person in control of the scene. That person should not dictate to others how to do their jobs, but rather should coordinate the efforts of the different response elements. He argued that each agency has to perform the tasks necessary to accomplish their goals, but they must do this from within the command structure.[15] As discussed in the preceding chapter, ICS was implemented at the Murrah scene in an effort to achieve the type of coordinated response recommended by Ricks.

The most significant command challenges were related to the crime scene aspect of the incident. As was discussed earlier, this was the first time that a disaster of this magnitude had been caused by an act of terror in this country and the variety of different agencies mobilized in support of the response generated some initial difficulties. Fire Chief Gary Marrs explained that coordinating the actions of the agencies might be challenging because in rescue and recovery you are digging through the rubble and pulling it out of the way trying to get to the people. In contrast, in a crime scene the emphasis is on not moving things.[16] This coordination challenge is relevant to this section of the book in that it shows the conflict across the tasks and goals being pursued by the response agencies. This is not to say that the police department and FBI were not interested in saving lives. However, their primary bureaucratic goals differed from those of the fire department. Gary Marrs felt that this was a problem in that the response community at the local level did "not have a section in their emergency response plan for doing rescue and recovery in the middle of a crime scene. So that was a learning process for [them] from the very beginning."[17] His sentiment was echoed by the ODCEM's AAR, which listed among its lessons learned: "state and local plans and exercises need to be changed to incorporate response forces working in and around a crime scene."[18]

Setting up a command post and determining its location is an important early task that must be undertaken by the commanders. A command post provides a centralized location where all response agencies can converge to coordinate their efforts and communicate. The AAR for the OCPD emphasized the importance of the location of the command post. It said that careful consideration should be given to locating a site that is accessible, but safe; that is close in proximity to the disaster, but outside the perimeter so that it does not interfere with operations; and that is secure but accessible to all agencies involved in the incident.[19] The command post for the Murrah incident was moved twice in the early stages of the response effort. The first command post location (NW 6th and Harvey) was deemed to be unsafe and too close to the blast site after the bomb scare at 10:30 a.m.

Police Chief Sam Gonzales indicated that before 10:30 there were too many resources too close to the incident. If there had been a second device (which terrorist sometimes use to target first responders), it could have caused real problems for the response.[20] Gonzales felt that this was a failure on the part of the response decision-makers to think about a secondary device or response dangers.[21] In part, this may be due to the fact that many of the responders who were first on the scene were thinking this was an accidental explosion rather than a terrorist incident.[22] The command post was moved again at 11:00 a.m. when Southwestern Bell made their facilities at One Bell Center on Harvey available for the responders' use.

In addition to establishing a command post, personnel from the Air Force and the Oklahoma National Guard had to be integrated with the response activities. Technically, the responders from the Air Force cannot be placed under the command of civil authorities. Similarly, the National Guard technically must be under the command of the Governor. However, neither distinction seems to have been a significant problem for direction and control. Both entities provided support to the response efforts in the form of personnel and materials.[23] The TAFB participated in that the base itself mobilized resources and personnel to the scene and in that the TAFB Fire Department had a mutual aid agreement with the city and participated in the search and rescue mission. Primarily the difficulties encountered with the Air Force and National Guard were associated with the magnitude of the disaster which compelled many to respond in ways they never had before. Adjutant General Steven Cortright said that the National Guardsmen were not trained for crime scene protection at the time of the disaster. They were involved in every phase of operations in providing three rings of security as well as preserving the crime scene.[24] He said that this was a challenge in that they were trying to develop methods for dealing with an unplanned response activity as the response itself unfolded, which caused confusion.[25] Fire Chief Proctor of the TAFB Fire Department indicated that the tasks he performed were consistent with what he would expect to have to do in a major disaster. However, at the same time he felt that he was unprepared for the magnitude of the impact of the things that he had to do.[26]

According to Steve Ferreira of TAFB, there were problems encountered with some of the TAFB personnel who were on-scene. Everyone wanted to help and it was difficult to get a handle on those who self-responded. Furthermore, there was limited accountability for those personnel who were on-scene who were not ordered there.[27]

Members of local government provided support to the response effort. Mayor Norick and Assistant City Manager Bullard each stated that the approach of both local executive offices was to stay out of the way and let the professionals deal with the crisis. Representatives of both offices actively participated in the response by notifying proper authorities about the disaster and by attempting to arrange to have needed resources sent to the scene. Both offices fielded a large volume of phone calls from people throughout the country who were seeking to give help or get information. Local government members also used the media to notify people to stay out of the downtown area.[28] Mayor Norick said that if it had been an exercise, he might have been asking more questions. As it was, he knew that he should be available so that he could provide the city's support where it was needed. He felt that his role was to provide information and represent the city, while making sure that firefighters and police knew he was around and cared about the progress of the response.[29] Response tasks for Governor Keating seemed to have been similar. According to Rick Buchanan, his Press Secretary at the time of the incident, Governor Keating's job was 10 percent media related. The other 90 percent was talking to first responders when they came off shift and dealing with those who wanted to help in the response.[30]

Utility companies also seemed to integrate relatively easily with the incident command. Tasks performed by the utility companies included shutting off the water to the Murrah Building to prevent flooding, providing supplies that were needed for the response effort (including barricades, heavy equipment, pumps, hard hats, respirators, torches, and chainsaws), fixing broken gas lines and damaged meters, erecting new power poles, installing floodlights, and running electrical and telephone lines to the command post and buildings.[31] There were no indications that there were difficulties in direction and command with the utility companies.

Communications

As was discussed previously, effective communications are a key to effective response. However, communications were very difficult in the first hours of this response as demand for systems exceed their capacity. Evidence of this fact is found in the spontaneous mention of communications as a problem by almost 60 percent of respondents. Planners must consider the lack of effective communications in the early phases of disaster response to be a given and determine what could be done in its absence to ensure effective response.[32] Every participating agency and private sector organization commented on the communications difficulty.

Different strategies were employed in attempts to get around the communications difficulties. First, it was decided that all of the commanders would have their command posts in the same location so that they could more easily disseminate information. Typical comments on this decision emphasized that communications failures necessitated this decision. Mike Grimes stated that "it was a real advantage to be able to walk next door and talk to them and talk to Oklahoma City Fire Department and Police Department."[33] EMSA's AAR emphasized that a lack of common frequency among agencies made it "imperative that each response agency keep a representative in the incident command area to communicate important information."[34]

Another strategy employed was the use of runners. This is a woefully inefficient means of communication. It requires a large number of individuals to be disengaged from other response tasks to simply run and deliver messages between responders. Furthermore, the potential exists for the runners to get caught up in the response and messages will not be delivered. Mike Murphy of EMSA stated that he was unaware of what was happening on the south side of the building because units kept disappearing and runners did not come back when they were sent toward the south of the building. He said that it was about 25 minutes before he realized there was another triage station on the south side of the building.[35] However, when all other means of communication fail, runners do provide some opportunity to share information.

Rick Buchanan, Governor Keating's Press Secretary, and Ann Burkle, Clinical Coordinator for St. Anthony's Emergency ward, commented that they relied on the media to get the vast majority of their information.[36]

Finally, at about 6:00 p.m. a Cellular-on-Wheels (COW) unit was operational on the scene. COW are mobile cellular antennas that can be utilized to boost the capacity of a cellular system when there is high demand. Southwestern Bell and AT&T both participated in the provision of COW and in the distribution of cellular phones to responders. The cell phones that were distributed to the responders had priority service, so that if they called and there was not an open line, a person who was on a nonpriority phone would have their call disconnected to free up the line. The phone number of each responder was recorded on a master list so that a dispatcher would know everyone's phone number. Responders were not permitted to trade or change phones in the interest of maintaining the integrity of the call list.[37]

Emergency Public Information

The provision of information to the general public about the incident and the response is essential to maintaining a calm and functional society. "Crisis research and case studies show that accurate, consistent, and expedited information calms anxieties and reduces problematic public responses such as panic and spontaneous evacuations."[38]

Billy Penn of FEMA said that providing information to the public after a disaster is similar to providing food, water, and shelter. He felt that without information there is unrest but with information comes relative peace.[39] Informing the public can also help to deter individuals from converging on the scene and direct them to locations where they can be reunited with their loved ones. Responsibility for providing such information should be established in advance of a major incident. Furthermore, FEMA recommends that plans incorporate a means for sharing and coordinating information, development and production of information materials, dissemination of information across many types of media, and monitoring of the news media to identify and rectify the problems.[40]

As discussed previously, the task of informing the public was carried out largely through the media. Tasks associated with this support function included determining where to position the media and how to engage them effectively in the response. The media arrived quickly and in massive numbers. One of the first challenges was determining where the media should be positioned so that they would be close enough to do their jobs, but so close that they would endanger themselves or hinder the responders. The OCPD's AAR stated that media from throughout the country converged on the city within minutes. This necessitated selection of a central media assembly point that was for media personnel only and was secured by the police throughout the incident.[41]

Among those who participated in this emergency support function were Jon Hansen (PIO, OCFD), Bill Citty (PIO, ODPD), Dan Stockton (PIO, Oklahoma Highway Patrol), Mayor Norick, Rick Buchanan (Governor Keating's Press Secretary), and Joevan Bullard (Assistant City Manager). Opinions were consistent that a preexisting positive relationship with the local media facilitated the provision of information. Jon Hansen felt that the response "dispelled the myth that public officials and media cannot work together in time of disaster."[42]

Members of the local response community tried to cater to their local media representatives. They said that they knew that they would have to continue to work with the members of the local media in the future and it was in the long-term interest of the community to maintain positive relations.[43] The local media were required to share all information they received with the national and international media, but most information was given to the local media first.[44]

The responders determined very early that they would coordinate information so that each person speaking to the media was providing consistent information. To be certain that the public was receiving consistent information, the responders updated key statistics and information on a white board inside the command trailer on the scene. Those who were going to speak to the media would check the board before doing so to get consistent, recent data.[45]

Members of the response team attempted to be accommodating of the media's needs in order to facilitate the provision of information. They had a motor home that was loaned to them and inside they had marker boards that detailed who the media were in the area and what their timelines were. The responders tried to make sure that they had people available for comments to meet media deadlines.[46] Mayor Ron Norick felt that a successful aspect of the response was that they started communicating with the media early in the response and were very up front about what they did not know.[47] He argued that they told them everything they could tell them and controlled information without hiding anything. According to Mayor Norick, they also tried to make sure that information came from credible people and that it was consistent.[48]

The media can be used to provide a great deal of information about the incident and the response to the general public. However, they can also be a source of challenges for the responders. Misinformation can be detrimental to response efforts, erode public confidence and create public unrest. There were some minor difficulties in dealing with the media that were experienced at the Murrah response. Ann Burkle recounted her experience where a little girl who had first aid experience showing up in the emergency room in response to a media report that told any person with medical experience to go to St. Anthony's Hospital.[49]

Two additional problems were experienced with the media. First responders learned during the incident not to tell the media they needed supplies of any kind. Many of those interviewed commented that someone went on television and said they needed gloves or batteries or some other supply and the scene would be inundated with whatever was requested.[50] Dan Stockton felt that one of his mistakes in the response was in telling the media that responders needed gloves. Stockton said that this resulted in people stopping by within an hour of his comment with cases and cases of gloves.[51] Finally, some of the members of the media were determined to cross into the inner perimeter to get pictures inside the building, which was strictly prohibited. Several individuals were arrested inside the perimeter, some of them dressed in emergency response uniforms, when it was found that they were members of the media.[52] There was one additional problem when a national network got images of the inside of the building and played them on national television. This was a potential threat to the ongoing good relations between the local media (who were upset that they were not given the same access as the nationals) and the emergency response community. Jon Hansen said that this was a major issue that was upsetting to both the responders and the media.[53]

Members of the media also faced unique challenges in performing their tasks. Sue Hale, Assistant Managing Editor of the *Daily Oklahoman*, was interviewed for this book. She said that the first challenge was in determining how they were going to cover the incident. The paper basically reorganized its staff so that everyone, regardless of their normal tasks, was assigned to covering a specific element of the incident and the response. In addition, the paper went into a 24-hour coverage mode that required the staff to work long hours under great pressure. This necessitated that the paper provide for the well-being of its employees, including feeding them and arranging for counselors to come in to help employees deal with the horrendous nature of the incident. Consistent with the views of the first responders, Hale felt that the positive relationship between the local media and the first response community facilitated the response.[54]

Protective Actions

For incidents such as the explosion in Oklahoma City, protective actions tasks include primarily the establishment of safe perimeters that prevent further injuries. Rapidly establishing a perimeter can facilitate command as it will stop the flow of individuals onto the scene and aid in establishing accountability. Typically, as it occurred in Oklahoma City, local and state law enforcement personnel dominated in this function. At the Murrah response, the National Guard was also instrumental in maintaining the perimeters.

An interesting divergence of opinion about the establishment of the perimeters occurred in the interviews. Some of the respondents indicated that they felt the perimeters were secured in a timely manner. Sam Gonzales and John Clark both commented that, with some exceptions, they felt the perimeters were set up relatively quickly and effectively.[55] Others felt that the slow rate with which the perimeters were set up hindered response. Gary Davis argued that tighter perimeter controls might have helped keep out civilians such as the nurse who was killed by falling debris as she tried to help on the scene. He felt that the lack of perimeter control in the first hour was among the least effective aspects of the response.[56] This is interesting in that it highlights the fact that determining success in many aspects of response is complicated by differing perspectives and expectations among the responders.

According to the official police department AAR, the inner perimeter, consisting of the block around the Murrah Building, was set at 10:40 a.m. or almost immediately after the second bomb scare cleared the building. An outer perimeter, which initially enclosed about twenty city blocks, was established at about 11:20.[57] Law enforcement personnel were more experienced in the establishment of perimeters and tended to think that they were set fairly rapidly. Firemen, who do not perform this task but whose jobs become easier once the perimeters are established, seemed to feel that it was not done efficiently. It may be that the expectation of the firemen that the perimeters be erected and effective more quickly was unrealistic.

One perimeter challenge that was not anticipated was the role that was played in collecting donated goods. People would simply drive up to the perimeter guards and thrust bags of supplies out their car windows. The challenge was then in determining how best to manage the goods so that they would be available when needed, but not in the way until then. Mike Grimes felt that one of the least effective aspects of the response in the first 12 hours was not knowing what to do with all the things people were driving up and giving to the responders. He said that the responders were just not ready for it.[58]

Another interesting phenomenon that occurred that was relevant to protective actions was that crime dropped markedly in the aftermath of the incident. This decrease in community conflict was not unanticipated. Dynes and Quarantelli noted that communities tend to be united by disaster situations because of the perception that the community is threatened as well as a development of values and norms that place a high priority on unity after a disaster.[59]

One of the major challenges of a response of this kind is maintaining normal operational capacities while devoting a significant amount of personnel to the incident. John Clark emphasized this in his interview. He indicated that a problem with major incidents is that you must respond with as little disruption to existing systems as possible. He said you have to protect the city and citizens, in addition to responding.[60] This staffing pressure was not a factor for protective actions in the early stages of the disaster due to the lack of routine crime in the city. Joevan Bullard said that they were all amazed by the lack of crime in the aftermath of the bombing. "We were so amazed that for several nights there wasn't a burglary... there wasn't a robbery... there wasn't a rape... Oklahoma City was virtually crime free for a period of time after the bombing because everybody was locked up on this."[61] Bullard also commented that there was no looting within the perimeter and crime dropped everywhere in the city in the weeks following the incident.[62]

Finally, it was relatively quickly recognized that there would be a need to establish a credentialing system. People continued to seek access to the site. With the many different agencies present, it was difficult for those on the perimeter to know who to let in and who to exclude. By the end of the 12-hour period, efforts were underway to develop an identification system that would be specific to the incident.[63] The police department's AAR indicated that adding the requirement for a credentialing system to plans may facilitate this aspect of response. "It is recommended that the fire and police create a standard access permit form which both agencies would readily recognize. Procedures for implementation, who issues the passes, and points of access need to be established very early in any future incident."[64]

Mass Care, Health, and Medical

For the purposes of this chapter, the responsibilities for mass care, health, and medical are combined. The section is structured in this way due to the fact that many of the same agencies will be participating in each of these support functions. To avoid duplication, they were combined. The section begins with the triage of those injured by the incident. It then proceeds through the movement of the survivors to medical facilities, caregiving at hospitals, and the handling of the deceased. It also includes the care for the families of victims and the mental health treatment for all involved in the incident and its response.[65]

Triage and Transport

Because of the nature of the disaster incidents, EMS issues and the transport of patients tend to be incredibly intense early in the incident. A total of 66 ambulances were utilized in the early phases of the response.[66] Once the initial onslaught of patients has been dealt with, medical personnel go into a support role. Mike Murphy of EMSA commented that their part in the response was over within 2–3 hours at which time they went into a supporting position for the rescue operation.[67] Gary Davis also commented that in multiple casualty incidents the emergency medical aspect of the response only lasts for 1–2 hours during which people either go to the hospital on their own or they are transported, they die or they are not hurt.[68]

Transport of patients from the scene to area hospitals began at approximately 9:27 a.m. (25 minutes after the explosion). By 10:00 a.m. it was determined that all of the immediate casualties had been taken to the hospital.[69] From that point on, EMSA and other EMS workers were on stand-by. It was believed that there might be voids in the rubble that would allow a person to survive for some time. There was hope that a pocket of survivors would be located and require immediate medical attention. To the disappointment of many responders, no pockets of survivors were ever found.

Almost immediately after the explosion, there was a massive influx of medical personnel on the scene. Jon Hansen felt that one of the failures of the response was that too many medical workers converged on the scene, emptying hospitals. He said that health care workers ran down the street with equipment taken from hospitals and emergency rooms to try and give assistance.[70] This complicated the triage process on the scene as the injured began to cluster around the many different people offering medical assistance and triage stations multiplied. The challenge was in tracking patients and ensuring accountability of the medical process. Mike Murphy said that at times there were up to ten spontaneous triage areas and that, if there had been a second wave of patients, they would not have been ready to control them.[71]

Another factor that challenges the provision of medical services is the tendency for the wounded to leave the scene to seek medical attention. This occurs for a number of different reasons including the following: people do not wish to wait for responders to have time to attend to them; well meaning citizens wish to help the wounded; some of the injured may object to the triage process and feel they can get help faster; and others feel that their condition is minor and either go to their personal physician or transport themselves to the hospital to free up ambulances and responders for those who are worse off.[72] The consequence of this behavior is an inability to track those impacted by the disaster, resulting in difficulties for family members who want to find them and problems for responders who want to account for those who were in the building. Mike Murphy said that emergency response plans say that you should triage all patients and then treat them accordingly, providing assistance on the scene to those with minor injuries and giving treatment to more severely injured patients in ambulances which serve to evenly distribute patients among the area hospitals. He said this looks good on paper, but the reality of this is that people will show up and transport the injured in their own private vehicles.[73] This was the case in Oklahoma City.

The transport of patients and the administration of first aid and triage is another area that is difficult to assess. Patients were transported to hospitals. However, only 210 of the 442 people treated in area hospitals were transported by emergency services personnel.[74] While this complicated the response community's task of tracking victims and reuniting them with their families, it is unlikely that anything could have been done about it. In fact, it may be the case that concerned citizens' actions saved some people. Given all that is going on in the time of a major disaster, it would be impossible for responders to monitor each person injured on the scene in the first hour of response to be certain no one was leaving the scene without their knowledge.

Challenges for Hospitals

Once patients arrived at the hospitals, medical personnel attended to their injuries. Given the influx of medical volunteers, there was no shortage of staff to help deal with the 442 people who were treated at 15 area hospitals.[75] According to Ann Burkle, Clinical Coordinator at St. Anthony's Hospital Emergency (where the majority of the patients were treated), there were so many medical personnel on hand that she was able to assign specialty physicians to each patient based on their injuries because they were all right there.[76] The victims suffered a variety of injuries, many of which were the result of the shrapnel that was generated when the façade of the building shattered.

As discussed previously, a great deal of the information the hospitals received about the nature and extent of the incident were through watching the television news media report on the incident. The communications problems complicated the medical response in several ways. First, it made it difficult for hospital personnel to get accurate information about how many injured were incoming and to which hospitals. Second, hospital employees were inundated with both family members and victims. Medical personnel were unable to communicate across hospital treatment areas and among the different hospitals to help reunite families more effectively. David Dagg, Director of Safety and Security at St. Anthony's Hospital, said that the hospital was deluged with victims and family members. He stated that the hospital was accepting patients at three locations and the lack of adequate communications between the hospital and the family assistance area made the hospital resort to runners, which was slow and inefficient.[77]

Another challenge for hospitals was in mobilizing sufficient supplies to deal with the abnormally large demand. In today's health care community, hospitals tend to keep only enough supplies on hand for 2–3 days of normal demand.[78] The rapid arrival of hundreds of victims with similar injuries quickly outstripped the supplies that were on hand. Fortunately, local medical supply vendors opened their warehouses and arrived on the scene relatively quickly with additional supplies.[79]

Maintaining the security of medical facilities is the final issue that will be discussed in this section. With the arrival of medical supplies and people in large numbers, ensuring the safety of medical facilities can become a challenge. At St. Anthony's a box was left in a driveway. No one knew what it contained and security was called because of concerns it may be a bomb. The hospital also began receiving bomb threats shortly after the incident.[80] Controlling the movement of people and dealing with potential threats is a challenge for which hospitals will have to prepare.

Challenges posed to the area hospitals were complex. Of course the obvious and primary challenge is to provide adequate treatment for the wide range of injuries that will be present. Furthermore, the hospitals needed to be able to track the patients they were treating from their initial intake to their discharge. While this is a task commonly performed by hospitals, it is rare that they have to process such a large number of patients in such a short period of time. In addition, the hospitals faced the difficulty of assisting families in finding their loved ones, which they do not have to do on a regular basis.[81]

Mental Health

While addressing the immediate medical and psychological needs of those injured in the incident is obviously an important task for responders, a less obvious requirement is tending to the mental health needs of the responders themselves. Responders to the Murrah Building saw the terrible devastation that was intentionally caused by another human being. The injuries suffered by some of the survivors and the deceased were gruesome and troubling to the responders. Furthermore, the fact that there were 19 children who were casualties of the attack was upsetting.

First responder communities are familiar with these issues and have instituted plans for helping the responders to deal with what they experience in the course of their jobs. Critical Incident Stress Debriefings (CISD) or Critical Incident Stress Management (CISM) systems are common techniques employed. CISD involves the counseling of responders in groups or individually as they leave the scene at the end of their shifts. For the police department, their CISD is formalized in the Cops Helping Alleviate Policemen's Problems (CHAPPS) program. Many of the response agencies required personnel to receive some sort of CISD. The "defusings" are mandatory because, as pointed out by Fire Chief Proctor of the TAFB Fire Department, many responders may need help whether they know it or not.[82] The City's official AAR, the AAR of the police department, and Gary Marrs emphasized how important the defusing process was in trying to counter the trauma and stress of participating in response in a difficult environment. Each commented that defusings began early in the response and were continued throughout.[83] The psychological impact of disasters such as this may, however, have an impact on private sector personnel for which there is no preestablished means for helping employees to cope.

Sue Hale of *The Daily Oklahoman* and Debby Hampton of the Red Cross both felt that taking care of their personnel was a unique challenge that they suffered in the response.[84] Sue Hale said that they had to find someone who could come in and provide counseling to the newspaper staff and reporters. She said that they had never had to do this before.[85] Similarly, Debby Hampton commented that she wished that the Red Cross had made their debriefings mandatory. She felt that the organization's turnover of about 90 percent in the aftermath of the bombing was a result of their personnel being psychologically impacted by their participation in the event.[86]

Addressing the psychological impact of disasters on the emergency response community, the private sector, and the community at large are important to the long-term health of the society itself in the aftermath of a major disaster. Given the tendency of the private sector and citizens to come together to give aid after an incident, it would be beneficial for communities to include tending to the long-term mental health of community members (beyond just the emergency responders themselves) in their emergency response plans.

Managing Casualties

There were 168 deaths associated with the Oklahoma City bombing. The death toll of the incident was in itself a challenge with which the response community had to cope. It was further complicated by the need to maintain a chain of evidence for each body due to the fact that the incident was a crime and because there was a need to identify each body for the families. Among the tasks that had to be performed in this area were determining when to start recovering the bodies, tracking the handling of each body from its location in the building to its final disposition, finding an appropriate place to store the bodies, and maintaining the chain of evidence for each body.[87]

The Medical Examiner's office worked with the FBI as the primary agents conducting these tasks. The Medical Examiner's office went from a normal operating staff of 36 to a staff of 300 in order to deal with the bodies.[88] Assistance in these tasks was also provided by the relatively new D-MORT teams that had been established through the FEMA. The D-MORT teams are comprised of relevant professionals including forensic pathologists, forensic investigators, dental identification specialists, and others.[89] These activities began during the first 12 hours of the response, but mostly functioned after this time period, so are only briefly mentioned here.

Assisting Families

Families were also victims of the crime. Many organizations played a role in helping the families to cope with this disaster. After the bombing, friends and families of those who worked at or near the Murrah Building, as well as those who thought someone they knew might have had business in the buildings that morning, were desperately trying to determine if their loved ones were alive. This was difficult because the disposition of most people from the buildings had yet to be established. Parents wandered the streets looking for their children. Others raced from hospital to hospital searching for their friends and family.

Shortly after the explosion it became clear that there needed to be a place for the families of those who were unaccounted for to gather and wait for news. This was a task that was not common to the responders. It was not part of the local response plans.[90]

Some disagreement exists about who initiated the creation of an FAC. It seems that initially there were two FAC set up, one by the American Red Cross and another by the Medical Examiner's office.[91] The American Red Cross became involved upon hearing it announced on television that people searching for children should report to there. It became their task to try and help parents locate the children who were at the nearby YMCA or in the America's Kids Day-Care Center at the Murrah Building. The Red Cross created a perimeter area where families could come and meet with medical health professionals and TVs that would provide them with some information about what was happening. Debby Hampton felt that this was the best that the Red Cross could do at that time and in this situation. She said that a lot has been improved in this aspect of Red Cross operations since then.[92]

Ray Blakenely believes that the reason for the confusion was that the agency responsible for establishing the FAC was never discussed in state and local plans.[93] He argued that this was something that had to be added to most emergency management plans. He felt that this was a weakness in their plans at the time.⁹⁴ By about 3:30 p.m., the FAC had been consolidated at the First Christian Church at NW 36th and N. Walker. From that point on, families could go to the FAC to wait for word about their loved ones. At the FAC, families were regularly briefed about the rescue operations. Ken Thompson, whose mother was among those killed in the incident, felt that the regular, honest briefings were helpful to the families. He said that the medical examiner's honesty "even to the point of hurting" was vital to the families because they knew that they were getting the truth.⁹⁵

Having the families together there also provided a single point where the Medical Examiner's office could collect information about those who were missing and determine how many people were in the building at the time of the explosion. When bodies were found in the building, their families were informed at the FAC before the information was released to the public. Counselors and clergy were on hand to comfort the families. Thompson argued that key components of an FAC would be mental health professionals, funeral directors, cellular service, clergy, food, regular reports from medical examiners, visible tokens of support from the community such as banners and other well-wishes, and a library of inspirational materials.⁹⁶

This chapter examined both bureaucratic structures and cultures. Framed by the FEMA ESF terminology, it focused on the tasks and goals performed by bureaucracies that participated in the response. The chapter highlighted the consistent and divergent opinions held by responders based on their role and culture. It also showed that conflict can occur when bureaucracies that are pursuing differing goals are compelled to work together during a major crisis. Finally, the chapter demonstrated that perspective, which is at least in part shaped by the bureaucracy one represents and its culture, colors one's perception of the relative success or failure of specific response actions.

Notes

1. See, for example, Paul Burstein, "Policy Domains: Organization, Culture, and Policy Outcomes," *Annual Review of Sociology* 17 (1991): 346.
2. Based on interviews and AAR.
3. FEMA, *Federal Response Plan*. 9230.1-PL (Washington, DC: FEMA, 2003); FEMA, *Managing the Emergency Consequences of Terrorist Incidents: Interim Planning Guide for State and Local Governments* (Washington, DC: FEMA, 2002), 21–31; and FEMA, *Guide for the Development of State and Local Emergency Operations Plans* (Washington, DC: FEMA, 1990).
4. Ashwood, interview by author.
5. William Waugh, "Current Policy and Implementation Issues in Disaster Preparedness," in *Managing Disaster: Strategies and Policy Perspectives*, Louise Comfort (ed.). (Durham, NC: Duke University Press, 1988), 114.
6. City of Oklahoma City, *Emergency Management Operations Plan*.
7. Marrs, interview by author.
8. Ashwood, interview by author.
9. Murphy, interview by author.
10. FEMA, *Managing the Emergency Consequences*, 21–24.
11. Clark, phone interview by author.
12. Gonzales, phone interview by author.

13. Grimes, interview by author.
14. FEMA, *Managing the Emergency Consequences*, 21–24.
15. Ricks, phone interview by author.
16. Marrs, Oklahoma City Documentation Team interview.
17. Ibid.
18. ODCEM, *After-Action Report*, 40.
19. OCPD, *After-Action Report*, 60.
20. Maniscalco and Christen, *Understanding Terrorism*, 48, 228, and 260.
21. Gonzales, phone interview by author.
22. This was commented on by many of those interviewed and is also included in some of the AAR.
23. Information was found in many sources including the City's AAR, National Guard AAR, Tinker AAR, and interviews with TAFB Fire Chief Proctor, Steve Ferreira of TAFB, Mayor Ron Norick, and Assistant City Manager Joevan Bullard, among others.
24. Comment from Adjutant General Steven Cortright in Oklahoma City National MIPT, *Oklahoma City*.
25. Adjutant General Steven Cortright, Oklahoma National Guard, *Memorandum After-Action Report Bombing, Alfred P. Murrah, Federal Building, Oklahoma City, Oklahoma* (1994).
26. Proctor, phone interview by author.
27. Ferreira, phone interview by author.
28. Norick and Bullard interviews with author.
29. Ron Norick, Mayor, City of Oklahoma City Documentation Team interview, May 25, 1995.
30. Rick Buchanan, Press Secretary to Governor Keating, phone interview by author, July 14, 2003.
31. Taken from City of Oklahoma City Document Management Team, *Final Report*; Greg Shirey and Allen Totten, Water Line Maintenance, City of Oklahoma City Documentation Team interview, August 24, 1995; and various other interviews.
32. FEMA, *Managing the Emergency Consequences*, 25–26.
33. Grimes, interview by author.
34. EMSA, *Terror in the Heartland*.
35. Murphy, interview by author.
36. Buchanan, phone interview by author; and Burkle, interview by author.
37. Marrs, interview by author.
38. FEMA, *Managing the Emergency Consequences*, 27.
39. Billy Penn, Public Affairs Officer, FEMA, phone interview by author, July 12, 2003.
40. FEMA, *Managing the Emergency Consequences*, 27.
41. OCPD, *After-Action Report*, 11.
42. Jon Hansen, "Working with the Media," *Fire Engineering: Special Issue: Oklahoma City Bombing, Volume 1*. October 1995.
43. Hansen, interview by author.
44. Ibid.; Stockton, interview by author; and Citty, interview by author.
45. Hansen, interview by author.
46. Marrs, "Report from Fire Chief."
47. Norick, City of Oklahoma City Documentation Team interview.

48. Norick, interview by author.
49. Burkle, interview by author.
50. See, for example, the interview with Hansen, Stockton, or Bullard, interview by author.
51. Stockton, interview by author.
52. Mike Shannon, Rescue Operations Chief, OCFD, interview by author, June 26, 2003.
53. Hansen, interview with author.
54. Hale, interview by author.
55. Gonzales, phone interview with author; and Clark, phone interview with author.
56. Davis, interview with author.
57. OCPD, *After-Action Report*, 8–9.
58. Grimes, interview by author.
59. D. Wenger, "Community Response to Disaster: Functional and Structural Alterations," citing Dynes and Quarantelli, (1970) "Property Norms and Looting: Their Patterns in Community Crisis," *Phylon* 31 (Summer): 168–182 in *Disasters: Theory and Research*, E. L. Quarantelli (ed.). Sage Studies in International Sociology (Thousand Oaks, CA: Sage Publications, 1978), 40.
60. Clark, phone interview by author.
61. Stockton, interview by author.
62. Joevan Bullard, Assistant City Manager, City of Oklahoma City Documentation Team interview, May 26, 1995.
63. Gonzales and Clark, phone interviews by author.
64. OCPD, *After-Action Report*, 61.
65. FEMA, *Managing the Emergency Consequences*, 28–30.
66. EMSA, *Terror in the Heartland*, 2.
67. Murphy, interview by author.
68. Davis, interview by author.
69. EMSA, *Terror in the Heartland*.
70. Hansen, interview by author.
71. Murphy, interview by author.
72. Waugh, "Co-ordination or Control."
73. Murphy, Oklahoma City Documentation Team interview.
74. Mark Robison, "EMS Treatment and Transport," *Fire Engineering: Special Issue: Oklahoma City Bombing, Volume 1* (October 1995); and EMSA, *Terror in the Heartland*, 4.
75. See, City of Oklahoma City Document Management Team, *Final Report*, 14; and Oklahoma State Department of Health, Injury Prevention Service, "Investigation of Physical Injuries Directly Associated with the Oklahoma City Bombing" (1995).
76. Burkle, interview by author.
77. Oklahoma City National MIPT, *Oklahoma City*. Interview in that document with David Dagg, Director of Safety and Security at St. Anthony Hospital.
78. Joseph Barbera, Anthony Macintyre, and Craig DeAtley, "Ambulances to Nowhere: America's Critical Shortfall in Medical Preparedness for Catastrophic Terrorism," in *Countering Terrorism: Dimensions of Preparedness*, Arnold Howitt and Robin Pangi (eds). (Cambridge, MA: MIT Press, 2003), 299–340.
79. Burkle, interview by author.
80. Ibid.
81. Ibid.

82. Proctor, phone interview by author.
83. City of Oklahoma City Document Management Team, *Final Report*, 157; OCPD, *After-Action Report*, 24; and Marrs, City of Oklahoma City Documentation Team interview.
84. Hale, interview by author.
85. Ibid.
86. Hampton, interview by author.
87. Ray Blakenely, Director of Operations, Oklahoma Medical Examiner's Office, phone interview by author, June 24, 2003.
88. Ibid.
89. Ibid.
90. Bullard, interview by author.
91. See interviews with Marrs, Blakenely, and Hampton.
92. Hampton, interview by author.
93. Blakenely, phone interview by author.
94. Bullard, interview by author.
95. Oklahoma City National MIPT, *Oklahoma City*.
96. Ibid.

CHAPTER SIX

Conclusions: Lessons Learned and Reinforced

The interviews conducted for this book provided rich and varied views about the relative success of the emergency response efforts, the influence of networks and bureaucratic culture, and the function of response bureaucracies. Interviews and other supporting materials were utilized to provide insights into four major research questions: (1) What was the nature and interrelationship of bureaucratic structures involved in the response? (2) Was there evidence of networking among the bureaucracies, and if so what was the impact of those networks? (3) Was bureaucratic culture present, and what was its impact? (4) Does the method employed by this book provide valuable information to evaluate the response and assist in planning for future incidents?

Responding to acts of terrorism is a process so unpredictable and challenging that it is unlikely that, as emphasized by William Penn of FEMA, there will ever be a "perfect" response.[1] There will always be variables for which the responders have not, and possibly could not have, planned. It is unlikely there will be any forewarning that a strike is imminent. The extent to which response will be judged effective will be based on the ability of the response community to save lives and move from chaos to a coordinated, interagency response.

As this project pursued many discrete research goals, the conclusions have been divided into three sections. First, the substantive conclusions will be presented. These are the findings about the emergency response itself and its consistency with the literature and response plans. These conclusions include reinforcement of arguments already in the literature as well as some new findings. This section also includes some information comparing U.S. response activities to those of other countries. The second set of conclusions will focus on the bureaucracy theory literature. Finally, the strengths and weaknesses of the methodology utilized in this book will be put forth.

Substantive Conclusions

This section provides analysis of the interviews regarding the emergency response actions taken at the scene. The approach will be to identify consistent and inconsistent points of view across the respondents to identify trends. Table 6.1 provides a summary of the frequencies of some of the important and recurring comments from the interviews and AAR.

First, opinions are consistent among sources and respondents that emergency response entities must have an all-hazards response plan and they must exercise it often. These exercises should consist of those people who will be the actual decision-makers in the event of a disaster. They should include people from all segments of the community (to include the private sector) who may be called upon to give assistance. These likely include but are not limited to those listed on Table 6.2. All of these entities need to be brought together to exercise the plans for several important reasons. It familiarizes them with the plans. It helps them to identify and learn what roles they are expected to play according to those plans.

Table 6.1: Summary of Comment Frequencies*

Comment	Valid Percentage
Tasks corresponded well or were somewhat related to expectations	78.6
Of those who identified they were using FEMA guidelines, felt that response activities complied fully or somewhat to FEMA guidelines	77.0
Responded that they did not think there was anything so unique about the incident response that it could not be learned from by other communities	73.1
Spontaneously mentioned the outpouring of volunteers and donations as a complicating factor for response	66.7
Spontaneously mentioned the second bomb scare as important to the response effort	60.6
Spontaneously mentioned communications as a response problem or listed it as one of the negative aspects of the response	60.6
To respondent's recollection, agency was utilizing FEMA guidelines for state and local emergency response planning	53.8
Spontaneously mentioned the importance of networking on effective response	51.5
Spontaneously mentioned chaos as a complicating factor for the response	51.5
Spontaneously mentioned the ICS as having positively impacted the response	45.5
Spontaneously mentioned the Emmitsburg training as having a positive impact on the response	39.4

*Comments and percents based on author's interviews.

Table 6.2: Minimal Representation at Major Multiagency Training Exercises*

Coroner's or Medical Examiner's Office	National Guard
DHS representatives	Police department
EMS	Potentially relevant private sector firms such as phone services, heavy construction, food service, etc.
Fire department	Private sector emergency support such as American Red Cross and Salvation Army
Highway Patrol	Public works and streets departments
Hospitals	Public health agencies
Local military (if relevant)	Representatives for mutual aid agreements
Local representatives of FBI, BATF, FEMA, and other national agencies	Representation for special response teams such as Search and Rescue, Emergency Response Teams, CISD, Hazmat, etc.
Media	State, county, and local emergency management agencies
Members of state and local emergency operations centers	Utilities
Members of state and local government	

*List generated by author based on interviews and observations.

Finally, and quite probably most importantly, such gatherings perform an essential function for effective response: they get the people who will have to work together in the same room so that they can network.

They start to get to know one another on a professional basis and perhaps begin the process of learning the roles others will be playing and the goals they will be pursuing. More frequent exercises will allow for new members of the response community to begin engaging in this networking process.

Several of the respondents commented that this aspect of the trip to Emmitsburg the year previous to the bombing had the greatest impact on the relative success of the response. As Fire Chief Marrs put it: "they need to talk back and forth . . . and a goodly percentage of that happened on coffee breaks and lunch breaks and around the beer tavern at night. I mean, that's where they got to know each other . . ."[2] For this reason, it may also be helpful to occasionally have the exercises out of the city in some other location. While this option may not be possible for each year's training due to cost, it provides the added benefit of getting the participants out of their normal working routine and home concerns. Because of their situation, participants are almost compelled to socialize—they cannot go home to their families at night or to a luncheon meeting in the afternoon. This kind of informal interaction before a crisis builds the trust and working relationships required for a more successful emergency response during a disaster.

Training and planning cannot possibly encompass all of the things that responders may encounter on the scene. However, training and planning help to get all the responders on the same page as to what their tasks and goals must be. This will help to guide the improvisation that will be required of the responders toward a commonly shared vision. Training may also help responders more quickly recover from the natural human emotions they will experience when they arrive on the scene.

The NRF, FEMA guidelines for state and local response planning, and the NIMS are a good starting point for response planners. The fact that 53.8 percent of respondents recalled that their organization was utilizing FEMA guidelines at the time of the incident provides evidence that local response entities are attentive to those guidelines. This is particularly insightful given that about 20 percent of the respondents were in positions (such as the media) where their daily tasks would not cause them to be exposed to the FEMA guidelines. Furthermore, the fact that 77 percent of those who recalled they were working with FEMA guidelines felt that activities fully or somewhat complied with them is evidence that the communities are not paying token compliance to the guidelines but are actually using them to help guide response.

Table 6.3 provides a comparison of the FEMA guidelines for state and local emergency response plans, the City of Oklahoma City Emergency Operations Plan, and the actions taken during the response. The response actions column is based on the interviews conducted and the AARs. The table shows the ways in which the response was consistent with the emergency operations plan as well as highlighting the areas in which the response diverged from the plan.

The table demonstrates that the response was largely consistent with the plans that were in place at the time of the incident. First, the actors seem to have taken on the tasks that were assigned to them in the response plan. The actions taken were fairly consistent with the plan, with some deviations (marked in italics on the table).

Under Direction and Control, it seems that the roles that the police and FBI would play were not anticipated due to the fact that the local plan did not really address a response that would be both a disaster and a crime scene. Also, the off-site Emergency Operations Center

Table 6.3: Evaluation of ESFs, Oklahoma City Emergency Operations Plans, and Response Actions Taken on April 19, 1995 in Oklahoma City (Key and Sources)

ESF[a]	Oklahoma City Emergency Operations Plan Annex[b]	Actions Detailed in Plan[c]	Actions Taken April 19, 1995[d]
Key	▪ Primary responsibility • Supporting agencies	▪ Taken from Basic Plan and relevant Annex (Phases of Plan, Response section)	▪ Regular text indicates things that were consistent with plan ▪ *Italics indicates things differing from the plan* ▪ **Bold indicates things not relevant to study**
Direction and Control	Annex A: Direction and Control • City Manager • Senior Fire Department official at the scene	▪ Develop the EOC ▪ Provide communications capability ▪ Assign qualified people to EOC ▪ Develop plans for EOC to operate on location ▪ Identify commanders ▪ Implement ICS	▪ Established by the Fire Chief, Police Chief, and FBI ▪ *FBI role necessitated due to crime scene* ▪ Implemented ICS ▪ Established command posts—forward in the building loading dock and SW Bell building ▪ *State EOC not used due to lack of effective communications*
Communications	Annex B: Communications ▪ **Emergency Management Staff** ▪ **EOC**	▪ Establish an EOC message center ▪ Ensure messages routed through message center ▪ Supervise EOC communications operations ▪ Implement disaster call list	▪ *No information found to say these actions were taken* ▪ *Most communications failed in first day of response* ▪ *Some radio function* ▪ *Limited cell phone use*
Emergency Public Information	Annex D: Emergency Public Information ▪ PIO	▪ Release public information ▪ Coordinate rumor control ▪ Schedule news conferences ▪ Assign PIO to manage	▪ Incident PIO assigned from fire department ▪ *Other agencies also had PIOs* ▪ Released public information ▪ Tried to assure consistency of information to offset rumors ▪ Scheduled news conferences

Table 6.3: Continued

ESF[a]	Oklahoma City Emergency Operations Plan Annex[b]	Actions Detailed in Plan[c]	Actions Taken April 19, 1995[d]
Protective Actions	Annex I: Law Enforcement ■ Chief of Police • Police department • Oklahoma County's Sheriff's Office • Oklahoma State Highway Patrol • Public Works • Mutual Aid Police Departments • Oklahoma State National Guard (if requested) • Oklahoma State Bureau of Investigation	■ Maintain law and order ■ Provide security for facilities ■ Cordon off disaster area ■ Provide traffic and crowd control ■ Patrol evacuation routes ■ Take shelter as necessary ■ Control access to restricted areas ■ Coordinate movement of disabled vehicles	■ Police, Highway Patrol, and National Guard all worked the cordon area ■ Police worked to establish ingress and egress routes to the scene
Mass Care	Annex M: Human Services ■ Red Cross • Oklahoma Department of Human Services, Welfare Administrator • Salvation Army	■ Establish service distribution centers ■ Provide care, shelter, clothing, and services ■ Coordinate relief activities and personnel ■ Publish lodging and mass feeding locations ■ Provide information points for victims needing assistance	■ Red Cross and Salvation Army worked to provide mass care ■ Information points established ■ Myriad building used in this regard ■ *Creation of FAC not anticipated by plan* ■ **Publications of lodging and mass feeding came later in response**
Health and Medical	Annex H: Health and Medical • Director City-County Health Department • First rescue unit on scene • EMSA	■ Provide medical care for casualties ■ Coordinate disease control operations ■ Coordinate sanitation activities ■ Ensure availability of potable water ■ Coordinate environmental health	■ Medical care on-scene provided by fire department rescue and EMSA ■ *No evidence found that Director of City-County Health Department active early on* ■ *Hospitals provided most of the health care—not mentioned in plan* ■ **Disease control, environmental health, and sanitation become important in later days of response**

(*Continued*)

Table 6.3: Continued

ESF[a]	Oklahoma City Emergency Operations Plan Annex[b]	Actions Detailed in Plan[c]	Actions Taken April 19, 1995[d]
Resources Management	Annex N: Resource Management • A member of the Emergency Management Staff • The City of Oklahoma City • Oklahoma Civil Defense Agency • A Resources Officer, once one is designated	• Keep records of services, labor, hours, wages and supplies • Establish a labor pool • Establish a resource distribution center • Manage and distribute required resources • Secure additional resources	• *Resources arrived—some requested, some not—in huge numbers and were poorly coordinated* • **Other activities took place over the course of the first several days of the response**

USAR	Annex K: Fire and Rescue
	• Fire Chief
	  • Fire department
	  • Police department
	  • Water Resources Department
	  • Public Works
	  • Mutual Aid Fire Departments
		• Provide fire suppression
		• Conduct rescue operations as capable
		• Assist radiological defense measures
		• Conduct hazardous materials operations
			• Fire department engaged in fire suppression
			• Police and fire both participated in rescue operations
			• **No radiological issue present**
			• **No hazardous materials operations required**

[a] Federal Emergency Management Agency, *Federal Response Plan*. 9230.1-PL (Washington, DC: FEMA, 2003); Federal Emergency Management Agency, *Managing the Emergency Consequences of Terrorist Incidents: Interim Planning Guide for State and Local Governments* (Washington, DC: FEMA, 2002), 21–31; and Federal Emergency Management Agency. *Guide for the Development of State and Local Emergency Operations Plans* (Washington, DC: FEMA, 1990).
[b] Author's compilation of information from the Basic Plan and Annexes of the City of Oklahoma City (February 1994) *Emergency Operations Plan*.
[c] Ibid.
[d] Author's compilation of information from interviews and AARs.

(EOC) was not utilized in the response due to the poor communications and the desire of the key command actors in the response to be at the scene. Communications provisions seem to have diverged significantly from the plan, but this was largely due to the almost complete failure of normal communications methods.

Emergency Public Information was consistent, but there were actually numerous PIOs at the scene on that day. Each agency assigned a PIO. However, the fact that Jon Hansen of the OCFD was designated the head of public information makes this fairly consistent with the plan. It would have been impossible, given the duration of the response efforts, for only one person to serve as PIO.

Mass Care diverged from the plan only in that the plan did not anticipate the need to create an FAC. Other than that factor, the establishment of a distribution center, provision of shelter, and coordination of relief activities were consistent with the plan.

Health and Medical was an interesting area of the plan, which seemed to vary significantly. Medical care on the scene seemed consistent with the plan. However, care for individuals after they were transported to hospitals was not mentioned in the plan. Most of the care provided for the victims was in hospitals. Also, there was nothing found in the interviews or AAR that indicated that the Director of the City-County Health Department was active in the response, especially in its early phases. However, Health and Medical provision seems to have occurred relatively effectively in the incident.

Resources Management grew to fit the plan in the days following the incident. The first day of the response was not consistent with the plan. Management, storing, and distribution of resources were, as indicated from interview materials, ineffective early in the response. This may be seen as a failure in the response in that it was anticipated by the plan that resources management would be an important issue in the aftermath of an emergency, but the early response efforts failed to take this into account.

Finally, Protective Actions and USAR were consistent with the Emergency Operations Plan Annex.

Review of the table in its entirety results in an interesting observation. Two main factors seem to have impacted responders' utilization of the Emergency Operations Plan—both of which are rooted in the unique nature of the incident. First, the utilization of a more complex structure for Direction and Control was necessitated by the fact that this was an act of terrorism and not a natural disaster. Second, many of the other divergences from the plan could be attributed to the large-scale and long-term response that was required. Had the incident been resolved in a few hours, it is unlikely that Emergency Public Information, Mass Care, and Resources Management would have been major issues for the responders. Improvisation was required to deal with the magnitude of the disaster.

The NIMS, which combines many of the elements of the NRF with the ICS, was created by the Department of Homeland Security (DHS) to give all responders the same framework for emergency management.[3] The DHS itself resulted from a desire to become better prepared to respond to major terrorist attacks. The performance of the DHS after Hurricane Katrina indicates that it is not yet able to command the federal level response particularly effectively. Nor does it seem that coordination across federal levels of government has improved. In the future, as DHS develops an operational culture and performs additional training exercises across the levels of government, it may be able to provide more coordinated support to emergency responders at the local level. Also, centralization of training and planning at DHS should help it to provide more consistent advice to local bureaucracies.

However, each of the response mechanisms put forth by DHS is just a starting point. Response planners must be prepared to build upon the basic framework provided to make a plan that is focused on local requirements, capabilities, and resources. Several respondents had recommendations for planners as they attempt to develop response plans based on federal guidelines. Joevan Bullard said that while FEMA guidelines provide a good framework, it is not one-size-fits-all and has to be tailored to the local situation.[4] Ed Hill reinforced this view and argued that municipalities have to take the guidelines and "build from them and customize them."[5] Furthermore, the planners must realize that the guidelines are not so much a method of response as a conceptualization of tasks and responsibility assignments.

When utilized, the actions taken by the responders are not dictated by the plan. Instead, the coordination of the many different response mechanisms is facilitated by the plan.

Reliance on federal mechanisms for planning may, in itself, be problematic. As alluded to earlier, in emergency response planning there is a dynamic tension between centralization and decentralization. Centralization (i.e. decision-making and planning at the federal level or the top-down approach):

> assures adequate fiscal, technical, administrative and political capacities to address complex emergency management problems; and decentralization assures adequate flexibility, shorter logistical and communication lines, and greater potential for targeting responses where they are most needed.[6]

Traditionally rancorous relations across the local, state, and federal-level agencies relevant to emergency management may complicate the dissemination of NIMS. However, NIMS may provide some benefit in that if it is implemented by the majority of localities, each will be prepared for the same response process and utilizing the same terminology regarding response.

Planning efforts should include members of the responder community and private sector in the process to capture their views and expertise. Furthermore, their inclusion gives them "ownership," or a stake in the success of the plans which may improve the implementation of the plans during an incident. Plans also must be distributed widely, including response agencies, private sector participants, and members of government. Too frequently plans are made to respond to a government directive or to comply with a grant requirement, but then the plans are not distributed to those who will ultimately implement them. Albert Ashwood argued that one of the biggest problems they have is that elected officials who are making decisions have no idea what is in the plans or why they contain what they do.[7] Distributing the plan widely, practicing the plan with the broadest possible inclusion of potential participants, and frequently updating the plan to account for changes in the community or technology will improve the chances for successful response.

A second conclusion is that, while you cannot plan for chaos, you can plan for the first phases of any response to be chaotic. "Intrinsic to emergency management, uncertainty needs to be incorporated into the policy process as a fundamental component."[8] Because responding to major incidents will, as described above, have many unique and complex characteristics based on the weapon used, the exact location of the attack, the number of locations hit, and many other variables, it is almost certain that the first hours of any response will be characterized as chaotic. There will be too many people on the scene trying to help. There will be donated resources arriving at the scene for response teams to use. Communications will be difficult if not impossible (see the third conclusion for more discussion). Planners cannot anticipate everything that will be needed or exactly what will be encountered by the response teams. What they can do is recognize that those hours will be chaotic and plan strategies for bringing order out of that chaos. The ability to bring about order and respond to society's needs is crucial to the larger society's long-term well-being and support of government.

While arguably imperfect, one such strategy, may be the ICS.[9] A strategy that features flexibility and scalability should be an important factor in plans as it will allow the responders to utilize the same response plan regardless of the incident's characteristics. A fundamental reason why this kind of system is critical is that it allows responders to utilize the same system day-to-day that they will be using in major incidents. It is unrealistic to expect that responders would be efficient if they were to try to do something markedly different from their standard operating procedures merely because the incident to which they are responding is nonroutine. However, it is impractical to function day-to-day based on the characteristics of a major disaster. A scalable system can accommodate both small and large incidents without raising unrealistic expectations. In fact, the 9/11 Commission cited the implementation of ICS at the Pentagon as one of the reasons for the success of the response effort.[10]

Only 45.5 percent of those interviewed mentioned the ICS as having positively impacted the response. In the emergency response literature, there are comments that the ICS has been adopted more readily by fire departments and that other emergency response entities were not using it—something that has been changing since 9/11.[11] If this were the case in Oklahoma City in 1995, one would expect this percent to be closer to the 21.2 percent of respondents that were employed by the fire department. The higher percentage may indicate that there was more general acceptance of the ICS across agencies in Oklahoma City. This may be due to the Emmitsburg training, which taught the ICS to the City's training participants and therefore may have encouraged more agencies to adopt and implement the method. The DHS promulgated the ICS as part of the NRP. The NRP was designed to coordinate and guide the federal-level response, but also provides guidance to state and local responders and planners.[12]

The third major reinforcing conclusion is that, as predicted by the emergency response literature, communications will be unavailable or unreliable for the first hours of the response. This phenomenon was observed at the Oklahoma City bombing, on 9/11 in New York City and Washington, DC, and in Louisiana after Hurricane Katrina. In each case response communities found it difficult, if not impossible, to communicate.[13] In fact, in New York City on 9/11, communications were sporadic if not missing entirely: between inside the World Trade Center and outside; among NYPD and FDNY; and between commanders and their subordinates.[14] It would be impractical for a city to develop a communications infrastructure with the capacity required for such an incident. Paying for the day-to-day operation of this overcapacity would be a poor allocation of resources. Preparing to deal with the aftermath of a terrorist strike in the absence of reliable communications entails two major kinds of plans.

First, there needs to be a plan in place for how to communicate in the absence of such traditional means as cell phones, landline telephones, pagers, and radios. As occurred in Oklahoma City, this may take the form of runners who communicate face-to-face at each end of the required communication. If this is the plan established, however, it should be clear to the runners that their sole role in the response will be to relay messages—at the Oklahoma scene runners frequently disappeared as they became engaged in the response rather than carrying their messages to both ends. This may be a function that members of Community Emergency Response Teams (CERT) could be trained to perform. If they were not in recognizable emergency response uniforms, they may be less likely to be pulled into other response activities.

Planners also need to develop a strategy for reinstating communications as quickly as possible. Where and how to acquire priority cell phones and COWs should be included in plans. Some kind of public-private agreement should be prearranged so that it can be relied upon during a disaster response.

Another reinforcing conclusion that can be drawn is that people will respond to tragedies with resources and volunteerism for which cities must be prepared. This was predicted by the literature, and witnessed at Oklahoma City, New York City and Washington, DC, on 9/11, and after Hurricane Katrina. Such resources (both personnel and material) can be invaluable to the response effort, but only if they are collected in a central location and cataloged so that they are available when needed. State and local planners should be prepared for the requirement that they be able to collect, inventory, and distribute these resources effectively. Sue Hale, who now works for the American Red Cross as well as the *Daily Oklahoman*, had an interesting perspective on this issue. She recommended that localities be prepared to create a resources clearinghouse coordinated by phone, on-line, or both. People would be given contact information for the clearinghouse by the media and then would be able to use the system to see exactly what resources were needed by the responders and "pledge" specific quantities of those resources.[15] This scheme would allow resources to be received in controlled quantities.

Concomitantly, responders must be trained not to mention things they may need or that might aid the response when they are speaking to the media. In some cases, offhand remarks by responders produced amazing quantities of goods. The command structure should determine who should make those appeals and how they should be handled.

In the aftermath of 9/11, volunteer management is something that is being facilitated by the federal government. The DHS has instituted the Community Emergency Response Teams program, which is essentially a credentialing system through which volunteers can receive training to assist during a major disaster. While this program provides a valuable resource in that on-scene commanders will have a means of identifying trained volunteers, it does not solve the potential problem of volunteerism. Planners and first responders must be prepared for the fact that many volunteers (with and without credentials) will turn out to help with the response efforts. The Community Emergency Response Teams will be easily identifiable and responders will know what they have been trained to do. However, there will be many, many more people who will turn out about whom nothing will be known. Plans should incorporate some sort of staging area where volunteers can be told to report and then utilized once their intentions and skills have been identified.

Establishing a strong perimeter system will be crucial to the response effort. It is important in that it facilitates all other aspects of the response. Effective perimeters will help with the control of volunteers, with the collection and management of resources, with maintaining the safety of the responders, and many other aspects of the response. In Oklahoma City, the second bomb scare expedited this task. Because planners cannot count on a similar event at every response, one respondent recommended that plans incorporate a "command stand down" within the first 2 hours of a major incident response.[16] This would require that all response elements disengage from the response long enough to ensure accountability and to gain control of the scene. It may also allow for an assessment of the structural integrity of any damaged buildings. While this may be difficult to enforce (as it was even difficult to get some responders to leave the building when a bomb was thought to be present), it provides a means for the Incident Commander to learn exactly what personnel are on the scene, to remove volunteers from danger and to utilize his response resources more effectively and systematically.[17]

As mentioned previously, an interesting difference of opinion about the perimeters was found in the interviews. While police personnel seemed to indicate that the perimeters were established relatively quickly, members of the fire department seemed to feel they were set up much too slowly. It is likely this disagreement is the result of the differing tasks, expertise, and focus of the different professionals.

In Oklahoma City, New York City and Washington, DC, first responders turned out in large numbers (whether dispatched or not) and charged into the buildings to begin rescue operations. It is what first responders do. However, the dangers of such actions were tragically demonstrated on 9/11 when the World Trade Center towers fell, killing 403 responders on the scene.[18] Given the uniquely catastrophic nature of the incident, it is not clear that a command stand down could have had any impact on 9/11. However, regrouping in the aftermath of a major disaster to ensure accountability and assess the dangers posed to responders by damaged infrastructure is something that could be incorporated into emergency response plans.

There was also general agreement among the respondents and the literature that cities must plan to carry out emergency response efforts without significant assistance from the federal government for at least the first 12–24 hours. Cities must plan and assure that equipment, personnel, and materials will be available to support a massive response effort for up to 24 hours. The creation of mutual aid agreements, which helped the response in Oklahoma City providing an expanded trained response community, is a means through which local communities can broaden their response capabilities without a major expenditure.

For the medical community, there are several conclusions. First, plans for the transport and treatment of the injured must account for the fact that many people will not go through the triage process but will instead provide their own transportation to medical facilities. This element of the plan will be particularly critical if there is a chemical agent or other contaminant involved in the attack. The real risk is that the walking wounded will be a health hazard themselves due to their exposure to the agent. Plans should also acknowledge that families will be turning out in large numbers at area hospitals seeking their loved ones. Plans should include some discussion of how to track those in hospitals and in transport. A means for communicating among the hospitals to determine where patients have been taken would greatly aid in the provision of medical services and the calming of family members.

One of the challenges in dealing with the medical community is that it requires the cooperation of competing private sector entities. In Oklahoma City, the medical community has voluntarily entered into an agreement for major incident response. Ann Burkle said that one of the things that Oklahoma City has done since the incident is develop a Metro Emergency Response Center (MERC) that is in charge of coordinating supplies and personnel for area hospitals in a disaster.[19] Coordination of medical facilities through the Metro Emergency Response Center or some other voluntary association should be included in response plans.

Another conclusion is that the media can be a friend or a foe during disaster response operations. The response community is very conscious of the need to create constructive relations with the media. Many responder agencies have programs in which they train individuals to serve as PIOs during major incidents. In such instances, official statements about the incident are coordinated and disseminated by the PIO. Networking among the response community and the media is also facilitated by the PIO, as it provides a single individual for the media to contact rather than a bureaucratic agency. Including the media in training exercises will help them to be more constructively integrated in the response as well as providing an opportunity to network them with the relevant PIOs.[20]

In the course of the interviews it was found that an important and potentially unacknowledged aspect of maintaining a healthy and robust response community is the provision of recognition to those who contributed to the response. Several of those interviewed commented that their organization or personnel were not invited to or acknowledged during events held to show the city's thanks to the responders, including those that may be in the private sector.[21] While it may not be necessary to include such actions in a response plan, it is in the interest of the community and future response efforts that participants be given the opportunity to be recognized for their contributions. The disappointment and hurt feelings caused by a failure to acknowledge certain sectors of society could cause resentment within the community, resulting in less friendly relations and a less secure society in the future.

Finally, one aspect of the literature on the bureaucracy was found to be markedly absent from the emergency response community. Several authors commented that bureaucratic officials are self-interested entities and that they act in a manner inconsistent with the public interest.[22] This is not evident in the actions of emergency response, where the street-level bureaucrats risk their lives to help members of the community impacted by the disaster. It is likely that here the tendency for societies to come together in the aftermath of a major incident has more impact on the situation than the day-to-day operations of bureaucracies.

Because this book is based on a single case study, it did not arrive at a broad definition of successful response. It is clear, however, that successful responses will be those that minimize the additional loss of life and property within the community. Therefore, some very important indicators of success were discovered through the case study that may help to guide emergency responders and planners as they prepare for future attacks. Those findings and reinforced arguments are presented briefly below. Each of these is supported by the existing emergency response literature and the interview materials. They are divided among the strategic and tactical response findings.

Strategic Findings and Reinforcement

The federal government is incapable of responding immediately to local disasters, whether natural or man-made, in numbers large enough to make a significant impact on the response. Therefore, the responsibility for the early stages of response will fall to local and state responders.

Flexible plans and training exercises based on those plans can help to prepare the emergency community for the complex and uncertain environment of a terrorist attack response.

The stronger and more comprehensive the networks built among the response community and relevant response support entities, the more likely the response is to be effective. These networks should expand beyond the response agencies to include public utilities, private sector entities that may be relevant, and the media.

The better able the response community is to engage in a symbiotic relationship with the media (i.e. one in which the media is helping to inform the public and the responders are providing the media with the news they need), the more effective the response is likely to be. A positive relationship here may also improve the public's perception of the government's response to the disaster. Protecting this relationship during the disaster is also important for the future interactions during everyday response activities.

Coordination is a key concept for emergency response to a major incident. This includes coordination: among local-level response entities; across the federal levels of government; between the public and private sectors; and among planners, responders, and government in advance of an incident. In many ways, coordination is a means to integrate the unique cultures of bureaucracies and private sector actors that must work together in a major incident response.

The more rapidly the scene is brought out of chaos and placed under the command of the response community, the more effective the response is likely to be. The ICS is one means through which this may be accomplished. One of the benefits of ICS is that it allows for different response communities to "plug" into the system so that, while there is an overarching Incident Commander, the direction and control of each type of response bureaucracy is carried out primarily by a member of that bureaucratic culture.

Tactical Findings and Reinforcement

The more rapidly effective perimeters can be erected, the more successful the response is likely to be. Firefighters and police officials should work together to establish a better understanding of this challenge between the two bureaucracies.

The more effectively response agencies can communicate during the initial phases of response, when normal modes of communications have been compromised by the disaster, the more effective the response is likely to be. The same applies with communications with the public. Public information activities can either greatly help or seriously hinder other emergency response tasks. Sharing timely and consistent information with the public can go a long way to maintaining calm.

Effective management of donated resources and volunteers will likely lead to more effective response. A balance must be struck by the responders between allowing the citizens to satisfy their civic desire to help and the need to maintain control of the scene, to protect citizens from further harm, and the ability to manage the volunteers and donated goods.

The more quickly survivors can be triaged and transported to hospitals for treatment, the more successful the response is likely to be. This finding is interesting in that it may conflict with the desire to establish strong perimeters and effective command. It may also conflict with the desire to assure that each victim is triaged on the scene. Members of the public providing transport to victims complicates response in some ways, but in the end may save some lives.

Addressing the mental health needs of the first response community, private sector volunteers, survivors, and the families of victims is important to the future health of the society and the response community.

This book contributes to the emergency response literature in several ways.

First, it supports arguments that many problems faced by emergency responders are persistent across crises and should be anticipated by the responders. There were few major challenges faced by the Oklahoma City response community that were not predicted in the scholarly literature. Second, the book provides expert insights into the challenges of emergency response to a major crisis. Finally, it provides advice to planners in both the responders' and the author's words, as to what they should do to plan more effectively for major disaster response.

Comparison with Other Countries

The many unique characteristics of the U.S. government, its terrorism response history, and emergency response community as compared to other countries made it difficult to do a major cross-country comparison. There were simply too many variables across countries to make methodologically valid comparisons. However, it is useful to examine briefly the challenges and responses to terrorist acts carried out by other countries.

Israel has had a long and violent history of terrorism. They have created a highly coordinated response mechanism to address the threat and increase preparedness. The organization with primary responsibility for response is the Home Front Command,

> which is a joint venture of the fire, police, and civil defense. The Home Front Command has many companies, such as engineering, medical, rescue, and logistics. The Home Front Command is leading the way to terrorism incident pre-planning through maintenance of a model for emergency response, command structure, and logistical planning and support.[23]

This structural approach to coordinating emergency response helps to ensure interoperability. However, it may be very difficult to implement such a tightly coordinated approach in the United States due to our larger geography and system of federalism.

Because of its history, Israel has a heightened state of awareness and preparedness for terrorism. In fact, the United States and Canada have both sought response training from Israel. The United States is identifying useful Israeli practices, such as preparing citizens for the threat and potential impact of domestic terrorism and encouraging families to create home disaster kits, to improve homeland security.[24] Canada recently had a delegation visit Israel to learn from their experience. One of the key things the Canadians were interested in was an Israeli hospital model that is designed to create surge capacity in hospitals during emergencies.[25]

As with the United States, Canadian emergency response is primarily a local responsibility.[26] Consistent with the U.S. experience, networking in advance of a disaster and emergency response planning are seen as important to effective emergency response.[27] Finally, communications systems in Canada are also prone to failure during major disasters.[28]

The United Kingdom has established an Incident Scene Management procedure that delegates authority and prioritizes establishment of inner and outer perimeters (cordons). The system is specifically designed to stop freelancing and support accountability. The U.K. system also establishes prearranged liaisons between ambulance services, police, and fire officials. These liaisons attend each others' planning and preparation meetings so that everyone is networked in advance of a disaster. Common terminology has also been established to facilitate communications among agencies.[29] The United Kingdom also facilitates emergency public information by relying on the media. In the aftermath of the transportation bombings in London in July 2005, the media was utilized to pass information to the public about how to evacuate London, who to contact about casualties, school safety, and to stay out of London if they did not need to be there.[30]

In an additional comparison, one major difference between the United States and the European Union is that the European Union has had a much easier time developing an interoperable standard for communications, in spite of the fact that the process was begun at the same time in both areas. They did so by quickly picking a technology on which to focus their efforts and then working in tandem across Europe to develop standards for operations.[31]

Japan experienced many of the same challenges faced in Oklahoma City during the sarin nerve gas attack on the Tokyo subway in 1995. On arrival at the scene, first responders found a confusing and rapidly evolving scene. Triage was complicated by the fact that victims' symptoms changed quickly. The medical community also had difficulty dealing with the large number of patients who transported themselves to hospitals.[32] In the Kobe earthquake of

January 1995, Japan also experienced the challenge of dealing with huge numbers of volunteers who turned out to help and the outpouring of donated goods. Coordinating the efforts of volunteers and the disposition of donations was a significant difficulty.[33]

Crisis management in Japan also suffers from the same challenge in the United States in that the public is only supportive of such efforts in the immediate aftermath of a disaster. As the country moves away from the most recent disaster into the future, the public becomes less supportive of personnel and budgetary requirements for crisis management.[34]

A unique challenge for Japan is that their emergency response bureaucracy is very hierarchical. However, as in the United States, the first phases of emergency response are handled by local emergency response officials.[35] When a major disaster strikes, emergency responders must break from the hierarchical culture and local officials must eschew routine top-down decision-making to deal with the disaster. This is difficult for individuals who may fear blame if things go wrong and are not trained to make these kinds of decisions.[36]

As in the United States, one of the means that Japan has adopted to assist localities in dealing with major disasters is the creation of mutual aid agreements among neighboring cities.[37] Another major similarity is that after the 1995 nerve gas attack the Japanese recognized major interoperability and incident management problems in their first response bureaucracies. Their response to remedying this was to create an additional national-level agency that is responsible for coordinating the efforts of the emergency response community.[38]

Based on this brief comparison, it seems that the United States is experiencing the same challenges as other countries as it prepares to confront the threat of large-scale terrorism. It also is clear that in at least some cases, in spite of differing national cultures and governing structures, similar policies have been adopted to attempt to respond to the threat more effectively.

Theory Conclusions

Three theoretical questions were evaluated in this book based on the findings from the case study. Arguments about bureaucratic structure were analyzed. In comparing the top-down and bottom-up conceptualizations of bureaucracy, it was found that both are present in emergency response. In fact, it may be that the presence of both patterns of interaction poses a challenge to effective emergency response. First, it is clear that emergency response is bottom up in practice. Local-level responders must be prepared to respond to a terrorist-imposed disaster during its initial phases. Their actions and requests for assistance then typically move up the bureaucratic chain of command to the federal level. In contrast, funding and plans to facilitate improved response tend to move from the top down. The federal government has issued planning guidance for state and local response entities. Those communities must have plans consistent with that guidance to qualify for some forms of federal assistance and programs. However, planning also takes place at the local level. If states and localities submit federally consistent plans as a form of "token compliance" but then formulate different local-level response plans, this could create confusion and friction when the federal, state, and local actors have to cooperate in the response.

In regards to networking arguments, it is clear that networks were in place in the Oklahoma City emergency response community. Those networks seemed to be quite fluid and flexible, making them more consistent with issue network arguments than subgovernment relationships. Response agencies of different professions and across levels of government that may rarely come into contact with one another come together in crises to formulate and implement a response. The response agencies also coordinate with members of government and the private sector during major response efforts. Actors participating in the response will be determined by the nature and extent of the disaster. These networks are facilitated by training exercises that include all possible response actors. Training like this allows first responders with different backgrounds to get to know each other well in advance of an emergency.

Training helps to form an organizational culture within the response bureaucracies. This establishes standard operating procedures and common norms and values for performance. The formulation of such a culture can be beneficial or detrimental in the time of an emergency. Within an individual response agency, the culture can help to set standards and expectations of individual actors. However, when diverse agencies must come together and coordinate their actions, the different cultures present in each agency could come into conflict. An example of this would be the conflict between the cultural value of protecting the crime scene held by law enforcement officials versus the cultural norm of fire department officials to get into the scene and rescue as soon and as thoroughly as possible. This highlights the fact that responders from the many participating agencies will see the problems of response differently.

This book contributes to the theoretical literature on bureaucracies in that it tested the arguments against another case study. It found that the theoretical arguments are applicable to major crisis situations in which response bureaucracies are participating. By applying classic bureaucracy theories to emergency response, the book strengthened the theoretical arguments.

Methodological Conclusions

Finally, this book also tested a method for evaluating emergency response. There were several findings about the use of FEMA guidelines for state and local response planning to structure interviews with emergency responders. The framework provided by the guidelines was easily applied to the variety of first responders who were interviewed. The first set of questions, which were asked of all respondents, were useful in gaining understanding of their general impressions of the successful and unsuccessful aspects of the response, as well as providing information about their role in the incident. The more detailed questions that were focused on the ESFs of organizations provided valuable insights into the performance of specific tasks.

Overall, the questionnaire and interviews yielded a large quantity of quality information from the respondents. Utilizing this instrument, or one like it, over time to evaluate large-scale response efforts would provide emergency response planners with crucial insights into the complexities of response. If this were done, there might be sufficient information to arrive at some more general conclusions about response that could help improve emergency response more generally. It would also help provide a resource for identifying effective strategies for response.

It was discovered, however, that applying the questionnaire immediately after the incident is crucial to gaining accurate and complete information. The almost ten-year lag between the Oklahoma City bombing and the interviews generated problems of recollection as well as difficulties in determining when intervening events were being included in the response.

Had the interviews been conducted immediately, the responses would have been more tightly focused on the incident of interest.

A second means to improve the book would have been beneficial to have interviewed more responders. While there was representation from most response agencies and each level of government, a more comprehensive list of respondents would have provided additional insights into the challenges and successes of the day. Again, however, the long time lag between incident and interview complicated attempts to conduct additional interviews.

In future research efforts, it would be helpful to test this method again on additional emergency response efforts. Obviously, one case study does not prove that this method is valid. However, the book demonstrates that it is a useful approach to evaluating emergency response activities. While it did help to gain valuable insights into the Oklahoma City bombing response, it may not be equally applicable in other cases. Additional testing would help to demonstrate whether the method is more generally useful across a wide variety of major emergency response activities. It would also be interesting and useful to expand the international comparison in a cross-country comparison of emergency response challenges and practices.

Notes

1. Penn, phone interview by author.
2. Marrs, interview by author.
3. DHS, *National Incident Management System* (Washington, DC: U.S. Government Printing Office, 2004).
4. Bullard, interview by author.
5. Hill, interview by author.
6. Waugh, "Co-ordination or Control," 19.
7. Ashwood, interview by author.
8. Comfort, "Designing Policy for Action," 5.
9. For a complete discussion of ICS, see FEMA. "FEMA Independent Study Program: IS-195 Basic Incident Command System," available at: http://training.fema.gov/EMIWeb/IS/is195lst.asp.
10. National Commission on Terrorist Attacks upon the United States, *The 9/11 Commission Report* (New York: W.W. Norton, 2003), 314–316.
11. See, for example, Ronald Perry, "Incident Management Systems in Disaster Management," *Disaster Prevention and Management* 12, 5 (2003): 405–412.
12. U.S. DHS, *National Response Plan* (Washington, DC: U.S. Government Printing Office, 2004).
13. See, for example, *The 9/11 Commission Report*.
14. Kettl, "Contingent Coordination," 256–257.
15. Hale, interview by author.
16. Hansen, interview by author.
17. See ibid. and Shannon, interview by author.
18. *The 9/11 Commission Report*, 311.
19. Burkle, interview by author.
20. See Hansen, interview by author; Stockton, interview by author; Hale, interview by author; and Citty, interview by author.

21. See, for example, Murphy, interview by author. The issue was also mentioned by Bullard, interview by author.
22. See, for example, B. Guy Peters, "The Problem of Bureaucratic Government," *The Journal of Politics* 43, 1 (1981): 56–82; Sally Coleman, Jeffrey Brudney, and J. Edward Kellough, "Bureaucracy as a Representative Institution: Toward a Reconciliation of Bureaucratic Government and Democratic Theory," *American Journal of Political Science* 42, 3 (1998): 717; and Arjen Boin and Paul t'Hart, "Public Leadership in Times of Crisis: Mission Impossible," *Public Administration Review* 63, 5 (1998): 545–546.
23. Buck, *Preparing for Terrorism*, 31.
24. Mimi Hill, "Government Training Citizens in Civil Defense," *USA Today*. February 13, 2003.
25. Leora Frucht, "Canada to Learn Lessons from Israel in Improving Emergency Response," *Jewish Toronto.net* downloaded on August 22, 2005.
26. Henry Hightower and Michel Coutu, "Coordinating Emergency Management: A Canadian Example," in *Disaster Management in the U.S. and Canada*, Richard Sylves and William Waugh (eds). (Springfield, IL: Charles C. Thomas, 1996), 69.
27. Ibid., 92–94.
28. Ibid., 89–91.
29. Buck, *Preparing for Terrorism*, 19–20.
30. BBC News, "London Blasts: Emergency Advice," *BBCNewsOnline*. Downloaded from http://news.bbc.co.uk/go/pr/fr/-/hi/uk/4659817.stm on July 7, 2005.
31. Viktor Mayer-Schonberger, "Emergency Communications: The Quest for Interoperability in the United States and Europe," in *Countering Terrorism: Dimensions of Preparedness*, Arnold Howitt and Robyn Pangi (eds). (Cambridge, MA: MIT Press, 2003), 321–335.
32. Buck, *Preparing for Terrorism*, 21–24.
33. Akira Nakamura, "Preparing for the Inevitable: Japan's Ongoing Search for Best Crisis Management Practices," in *Managing Crises: Threats, Dilemmas, Opportunities*, Rosenthal, Boin, and Comfort (eds). (Springfield, IL: Charles C. Thomas Publisher, Ltd., 2001), 313.
34. Nakamura, "Preparing for the Inevitable," 309.
35. Robyn Pangi, "Consequence Management in the 1995 Sarin Attacks on the Japanese Subway System," in *Countering Terrorism: Dimensions of Preparedness*, Arnold Howitt and Robyn Pangi (eds). (Cambridge, MA: MIT Press, 2003), 379–381.
36. Nakamura, "Preparing for the Inevitable," 30.
37. Pangi, "Consequence Management," 380–381.
38. Ibid., 386–391.

Appendix A: Interviews Conducted

Person Interviewed	Position at Time of Incident
Ashwood, Albert	Director, ODCEM
Blakenely, Ray	Director of Operations, Medical Examiner's Office
Buchannan, Rick	Press Secretary for Governor Frank Keating
Bullard, Joevan	Assistant City Manager
Bunch, Kenneth	Assistant Chief, OCFD
Burkle, Ann	Clinical Coordinator, St. Anthony Hospital Emergency Department
Citty, Bill	Information Officer, Captain, OCPD
Clark, John	Lieutenant, OCPD Office of Emergency Management
Davis, Gary	Commander, Emergency Management Services for the OCFD
Farha, Dr. Bryan	Counselor and Professor—interview not used because he was not engaged in response within 12 hours of incident
Ferreira, Steve	Non-Commissioned Officer in Charge for Disaster Management, TAFB
Feurborn, Tom	ODCEM
Gonzales, Sam	Chief of the OCPD
Grimes, Mike	Captain, Oklahoma Highway Patrol
Hale, Sue	Assistant Managing Editor, The *Daily Oklahoman*
Hale, Troy	Plans, Operation, and Support, Oklahoma National Guard
Hampton, Debby	Local Volunteer Coordinator, American Red Cross
Hansen, Jon	Public Information Officer, OCFD
Hill, Ed	Emergency Response Team Captain, OCPD
Marrs, Gary	Incident Commander, Chief, OCFD
Moss, Ron	Rescue Operations Commander, OCFD
Murphy, Mike	Operations Officer, EMSA
Norick, Ron	Mayor, City of Oklahoma City
Penn, William (Billy)	Public Affairs Officer, FEMA
Proctor	Rescue Operations Team Captain, TAFB FD
Ricks, Bob	Special Agent in Charge, FBI
Shannon, Mike	Rescue Operation Chief, OCFD

Stockton, Dan	Public Information Officer, Lieutenant on the Oklahoma Highway Patrol
Storment, Steve	Assistant Chief, Phoenix Urban Search and Rescue Team—interview not used because he was not engaged in the response within 12 hours
Warren, Ronny	Patrol Sergeant and Emergency Medical Technician in the OCPD
Young, Cornelius	Major, OCFD

Appendix B: Interview Questionnaires

Questionnaire

Date   Time
Name of Respondent
Current Position

1. Job title at the time of the incident

2. Role in the incident

3. Would you categorize your position as representing:
   ☐ Federal government
   ☐ State government
   ☐ Local government
   ☐ Health care worker
   ☐ EMS
   ☐ Others, explain

4. Time of first arrival on the scene

5. Duration of time on the scene within the first 12 hours of response

6. What tasks did you perform during the incident response?

7. Would you say that the tasks you performed:
   ☐ Corresponded well with those tasks you are generally expected to perform in your job
   ☐ Were somewhat related to your normal job tasking
   ☐ Were markedly different from your normal job tasking
   ☐ Others

Please explain:

8. To the best of your recollection, in 1995 did the response guidelines for your organization comply with FEMA guidelines?   ☐ Yes   ☐ No

9. To what extent do you feel the response itself complied with FEMA-generated response guidelines?
   ☐ Fully complied
   ☐ Somewhat complied
   ☐ Did not comply
   ☐ Do not know

Comment or explain:

10. When (time and day) did the first federal agencies, aside from those housed in the Murrah building, arrive on the scene to assist with the response efforts?

11. What was (were) the most effective aspect of the response effort? Why was (were) that (those) aspect(s) most effective?

12. What was (were) the least effective aspect of the response effort? Why was (were) that (those) aspect(s) least effective?

13. What aspects of the response do you think were unique to the incident and would be unlikely to be observed elsewhere?

14. What aspects of the response do you think would provide good "lessons learned" for other cities and states as they prepare to respond to acts of terror?

Now we are going to talk about various aspects of the response effort. The questionnaire is structured across the FEMA for state and local governments found in *Managing the Emergency Consequences of Terrorist Incidents* in order to gauge the relative effectiveness of those categories. The format also allows me to gain information about the relative success of response in a variety of categories rather than simply for the overall response.

It is understood that you may not have experienced or witnessed the response activities in each category. I am interested in your views for only those responses in which you have firsthand experience. Furthermore, to the extent possible, I would like you to focus on the first 12 hours of the response effort.

Direction and Control

Local government emergency response organizations will respond to the incident scene(s) and make appropriate and rapid notifications to local and state authorities (Table 5.1). Control of the incident scene(s) most likely will be established by local first responders from either fire or police. To assure continuity of operations, it is important that the Incident Command Post be established at a safe location and at a distance appropriate for response to a suspected or known terrorist incident. In addition, in severe terrorist attacks, response operations may last for very long periods, and there may be more leadership casualties due to secondary or tertiary attacks or events. Planning should therefore provide for staffing key leadership positions in depth.

A.1 I understand that you are familiar with the direction and control efforts undertaken during the response. Is this information correct?
  Yes ☐ Proceed with following questions
  No ☐ Proceed to **communication** category

A.2 What were your experiences and observations about direction and control while you were participating in the response effort?

A.3 Overall, would you say that the direction and control was:
  ☐ a. Very effective
  ☐ b. Somewhat effective
  ☐ c. Effective
  ☐ d. Somewhat ineffective
  ☐ e. Ineffective
  ☐ f. Don't know

A.4 What aspects of direction and control were most effective?   Why?

A.5 What aspects of direction and control were least effective?   Why?

A.6 Were efforts undertaken during the response to alter direction and control?
  Yes ☐ Proceed with following questions
  No ☐ Proceed to **communications**

A.7 What changes were made to direction and control during the response?

A.8 Did those changes:
  ☐ a. Improve direction and control
  ☐ b. Hinder direction and control
  ☐ c. Have no impact on direction and control
  ☐ d. Don't know

A.9 Were there additional changes that could have been made to improve direction and control?

Communication

In the event of a WMD incident, rapid and secure communication is important to ensure a prompt and coordinated response. Strengthening communications among first responders, clinicians, emergency rooms, hospitals, mass care providers, and emergency management personnel must be given top priority in planning. Planning should include adding 9-1-1 resources when an event requires extraordinary response.

In addition, terrorist attacks have been shown to overload nondedicated telephone lines and cellular telephones. In these instances, the Internet has proven more reliable for making necessary communications connections, although it should be recognized that computers may be vulnerable to cyber attacks in the form of viruses. It is recommended that response organizations both establish relevant Internet connections with all coordinating emergency response organizations and have the use of these connections formalized in plans and practiced during training, drills, and exercises.

B.1 I understand that you are familiar with the communication efforts undertaken during the response. Is this information correct?
Yes ☐ Proceed with following questions
No ☐ Proceed to **warning** category

B.2 What were your experiences and observations about communications while you were participating in the response effort?

B.3 Overall, would you say that communications were:
☐ a. Very effective
☐ b. Somewhat effective
☐ c. Effective
☐ d. Somewhat ineffective
☐ e. Ineffective
☐ f. Don't know

B.4 What aspects of communications were most effective?   Why?

B.5 What aspects of communications were least effective?   Why?

B.6 Were efforts undertaken during the response to alter communications?
Yes ☐ Proceed with following questions
No ☐ Proceed to **warning**

B.7 What changes were made to communications during the response?
B.8 Did those changes:
☐ a. Improve communications
☐ b. Hinder communications
☐ c. Have no impact on communications
☐ d. Don't know

B.9 Were there additional changes that could have been made to improve communications?

Warning

Every incident is different. There may or may not be warning of a potential WMD incident. Factors involved range from intelligence gathered from various law enforcement or intelligence agency sources to an actual notification from the terrorist organization or individual. The Emergency Operations Plan should have HazMat facilities and transportation routes already mapped, along with emergency procedures necessary to respond.

C.1 I understand that you are familiar with the warning efforts undertaken during the response. Is this information correct?
Yes ☐ Proceed with following questions
No ☐ Proceed to **emergency public information** category

C.2 What were your experiences and observations about warning while you were participating in the response effort?

C.3 Overall, would you say that warnings were:
☐ a. Very effective
☐ b. Somewhat effective
☐ c. Effective
☐ d. Somewhat ineffective
☐ e. Ineffective
☐ f. Don't know

C.4 What aspects of warnings were most effective?   Why?
C.5 What aspects of warnings were least effective?   Why?
C.6 Were efforts undertaken during the response to alter warning?
Yes ☐ Proceed with following questions
No ☐ Proceed to **emergency public information**

C.7 What changes were made to warning during the response?

C.8 Did those changes:
☐ a. Improve warning
☐ b. Hinder warning
☐ c. Have no impact on warning
☐ d. Don't know

C.9 Were there additional changes that could have been made to improve warning?

Emergency Public Information

Terrorism is designed to be catastrophic. The intent of a terrorist attack is to cause maximum destruction of lives and property; create chaos, confusion, and public panic; and stress local, state, and federal response resources. Accurate and timely information, disseminated to the public and media immediately and often over the course of the response, is vital to minimize accomplishment of these terrorist objectives.

Crisis research and case studies show that accurate, consistent, and expedited information calms anxieties and reduces problematic public responses such as panic and spontaneous evacuations that terrorists hope will hamper response efforts.

D.1 I understand that you are familiar with the emergency public information efforts undertaken during the response. Is this information correct?
   Yes ☐ Proceed with following questions
   No ☐ Proceed to **protective action** category

D.2 What were your experiences and observations about communications while you were participating in the response effort?

D.3 Overall, would you say that communications were:
   ☐ a. Very effective
   ☐ b. Somewhat effective
   ☐ c. Effective
   ☐ d. Somewhat ineffective
   ☐ e. Ineffective
   ☐ f. Don't know

D.4 What aspects of emergency public information were most effective?   Why?

D.5 What aspects of emergency public information were least effective?   Why?

D.6 Were efforts undertaken during the response to alter emergency public information?
   Yes ☐ Proceed with following questions
   No ☐ Proceed to **protective action**

D.7 What changes were made to emergency public information during the response?

D.8 Did those changes:
   ☐ a. Improve emergency public information
   ☐ b. Hinder emergency public information
   ☐ c. Have no impact on emergency public information
   ☐ d. Don't know

D.9 Were there additional changes that could have been made to improve emergency public information?

Protective Actions

Evacuation may be required from inside the perimeter of the scene to guard against further casualties from contamination by primary release of a WMD agent, the possible release of additional WMD, secondary devices, or additional attacks targeting emergency responders. Temporary in-place sheltering may be appropriate if there is a short-duration release of hazardous materials or if it is determined to be safer for individuals to remain in place. Protection from biological threats may involve coercive or noncoercive protective actions, including isolation of individuals who pose an infection hazard, quarantine of affected locations, vaccination, use of masks by the public, closing of public transportation, limiting public gatherings, and limiting intercity travel. As with any emergency, state and local officials are primarily responsible for making protective action decisions affecting the public. Protocols should be established to ensure that important decisions are made by persons with the proper decision-making authority. The Terrorist Incident Appendix (TIA) should include provision for coordinating protective actions with other affected jurisdictions. Planning should also address ways of countering irrational public behavior that can hinder protective actions.

E.1 I understand that you are familiar with the protective actions efforts undertaken during the response. Is this information correct?
Yes ☐ Proceed with following questions
No ☐ Proceed to **mass care** category

E.2 What were your experiences and observations about protective actions while you were participating in the response effort?

E.3 Overall, would you say that protective actions were:
☐ a. Very effective
☐ b. Somewhat effective
☐ c. Effective
☐ d. Somewhat ineffective
☐ e. Ineffective
☐ f. Don't know

E.4 What aspects of protective actions were most effective?   Why?

E.5 What aspects of protective actions were least effective?   Why?

E.6 Were efforts undertaken during the response to alter protective actions?
Yes ☐ Proceed with following questions
No ☐ Proceed to **mass care**

E.7 What changes were made to protective actions during the response?

E.8 Did those changes:
☐ a. Improve protective actions
☐ b. Hinder protective actions
☐ c. Have no impact on protective actions
☐ d. Don't know

E.9 Were there additional changes that could have been made to improve protective actions?

Mass Care

The location of mass care facilities will be based partly on the hazard agent involved. Decontamination, if it is necessary, may need to precede sheltering and other needs of the victims to prevent further damage from the hazard agent to either the victims themselves or the care providers. The American Red Cross (the primary agency for mass care), the Department of Health and Human Services, and the Department of Veteran Affairs should be actively involved with the planning process to determine both in-place and mobile mass care systems for the TIA. A midpoint or intermediary station may be needed to move victims out of the way of immediate harm. This action would allow responders to provide critical attention (e.g., decontamination and medical services) and general lifesaving support, then evacuate victims to a mass care location for further attention. The following are general issues to consider for inclusion in the TIA:

a. Location, setup, and equipment for decontamination stations, if any.
b. Mobile triage support and qualified personnel.
c. Supplies and personnel to support in-place sheltering.
d. Evacuation to an intermediary location to provide decontamination and medical attention.
e. Determination of safety perimeters (based on agent).
f. Patient tracking/record keeping for augmentation of epidemiological services and support.

F.1 I understand that you are familiar with the mass care efforts undertaken during the response. Is this information correct?
  Yes ☐ Proceed with following questions
  No ☐ Proceed to **health and medical** category

F.2 What were your experiences and observations about mass care while you were participating in the response effort?

F.3 Overall, would you say that mass care was:
  ☐ a. Very effective
  ☐ b. Somewhat effective
  ☐ c. Effective
  ☐ d. Somewhat ineffective
  ☐ e. Ineffective
  ☐ f. Don't know

F.4 What aspects of mass care were most effective? Why?

F.5 What aspects of mass care were least effective? Why?

F.6 Were efforts undertaken during the response to alter mass care?
  Yes ☐ Proceed with following questions
  No ☐ Proceed to **health and medical**

F.7 What changes were made to mass care during the response?

F.8 Did those changes:
  ☐ a. Improve mass care
  ☐ b. Hinder mass care
  ☐ c. Have no impact on mass care
  ☐ d. Don't know

F.9 Were there additional changes that could have been made to improve mass care?

Health and Medical

The basic Emergency Operations Plan should already contain a Health and Medical Annex. Issues that may be different during a terrorist incident and that should be addressed in the TIA include decontamination, safety of victims and responders, in-place sheltering or quarantine versus evacuation, and multihazard/multiagent triage.

Planning should anticipate the need to handle large numbers of people who may or may not be contaminated but who are fearful about their medical well-being. In addition, the TIA should identify the locations and capacities of medical care facilities within the jurisdiction and in surrounding jurisdictions. The TIA should also include a description of the capabilities of these medical care facilities, especially with regard to trauma care. Depending on the nature and extent of a terrorist attack, the most appropriate medical care facility may not necessarily be the closest facility.

G.1 I understand that you are familiar with the health and medical efforts undertaken during the response. Is this information correct?
   Yes ☐ Proceed with following questions
   No ☐ Proceed to **resource management** category

G.2 What were your experiences and observations about health and medical tasks while you were participating in the response effort?

G.3 Overall, would you say that health and medical response was:
   ☐ a. Very effective
   ☐ b. Somewhat effective
   ☐ c. Effective
   ☐ d. Somewhat ineffective
   ☐ e. Ineffective
   ☐ f. Don't know

G.4 What aspects of health and medical response were most effective?   Why?

G.5 What aspects of health and medical response were least effective?   Why?

G.6 Were efforts undertaken during the response to alter health and medical response?
   Yes ☐ Proceed with following questions
   No ☐ Proceed to **resource management**

G.7 What changes were made to health and medical response during the response?

G.8 Did those changes:
   ☐ a. Improve health and medical response
   ☐ b. Hinder health and medical response
   ☐ c. Have no impact on health and medical response
   ☐ d. Don't know

G.9 Were there additional changes that could have been made to improve health and medical response?

Resources Management

The following considerations are highly relevant to WMD incidents and should be addressed, if appropriate, in one or more appendixes to a resource management annex:

a. Nuclear, biological, and chemical response resources that are available through interjurisdictional agreements (e.g., interstate pacts).
b. Unique resources that are available through state authorities (e.g., National Guard units).
c. Unique resources that are available to state and local jurisdictions through federal authorities (e.g., the National Pharmaceutical Stockpile, a national asset providing delivery of antibiotics, antidotes, and medical supplies to the scene of a WMD incident).
d. Unique expertise that may be available through academic, research, or private organizations.
e. Trained and untrained volunteer resources and unsolicited donated goods that could arrive at the scene.

H.1 I understand that you are familiar with the resource management efforts undertaken during the response. Is this information correct?
Yes ☐ Proceed with following questions
No ☐ Proceed to **urban search and rescue** category

H.2 What were your experiences and observations about resources management while you were participating in the response effort?

H.3 Overall, would you say that resources management was:
☐ a. Very effective
☐ b. Somewhat effective
☐ c. Effective
☐ d. Somewhat ineffective
☐ e. Ineffective
☐ f. Don't know

H.4 What aspects of resources management were most effective?   Why?

H.5 What aspects of resources management were least effective?   Why?

H.6 Were efforts undertaken during the response to alter resources management?
Yes ☐ Proceed with following questions
No ☐ Proceed to **urban search and rescue**

H.7 What changes were made to resources management during the response?

H.8 Did those changes:
☐ a. Improve resources management
☐ b. Hinder resources management
☐ c. Have no impact on resources management
☐ d. Don't know

H.9 Were there additional changes that could have been made to improve resources management?

Urban Search and Rescue

Urban Search and Rescue (USAR), Emergency Support Function (ESF) #9 in the Federal Response Plan, involves rapid deployment of USAR task forces to provide specialized life-saving assistance to state and local authorities, including locating, extricating, and providing on-site medical treatment to those trapped in collapsed structures. FEMA is the agency with primary responsibility for this ESF. FEMA's Emergency Management Institute (EMI) and the National Fire Academy are integrating USAR more closely into their curriculums.

I.1 I understand that you are familiar with the urban search and rescue efforts undertaken during the response. Is this information correct?
Yes ☐ Proceed with following questions
No ☐ Proceed to **final questions** category

I.2 What were your experiences and observations about urban search and rescue while you were participating in the response effort?

I.3 Overall, would you say that urban search and rescue was:
☐ a. Very effective
☐ b. Somewhat effective
☐ c. Effective
☐ d. Somewhat ineffective
☐ e. Ineffective
☐ f. Don't know

I.4 What aspects of urban search and rescue were most effective?   Why?

I.5 What aspects of urban search and rescue were least effective?   Why?

I.6 Were efforts undertaken during the response to alter urban search and rescue?
Yes ☐ Proceed with following questions
No ☐ Proceed to **final questions**

I.7 What changes were made to urban search and rescue during the response?

I.8 Did those changes:
☐ a. Improve urban search and rescue
☐ b. Hinder urban search and rescue
☐ c. Have no impact on urban search and rescue
☐ d. Don't know

I.9 Were there additional changes that could have been made to improve urban search and rescue?

Final Questions

Based on your experiences with this incident, do you feel that the FEMA guidance categories are adequate for helping states and localities engage in response planning?
☐ Yes
☐ No

Why?

How would you alter them, if at all?

Bibliography

Advisory Panel to Assess Domestic Response Capabilities for Terrorism Involving Weapons of Mass Destruction. *Annual Report to the President and Congress: II. Toward a National Strategy for Combating Terrorism.* Arlington, VA: RAND Corporation, 2000.
———. *Annual Report to the President and Congress: III. For Ray Downey.* Arlington, VA: RAND Corporation, 2001.
———. *First Annual Report to the President and Congress: Assessing the Threat.* Arlington, VA: RAND Corporation, 1999.
Agranoff, Robert and Michael McGuire. "American Federalism and the Search for Models of Management," *Public Administration Review* 61, 6 (2001): 650–660.
Anonymous. "Nichols Sentenced," *The Daily Oklahoman*, August 9, 2004, via http://bombing.newsok.com/bombing/history/ on November 20, 2008.
Ashwood, Albert, Program Manager, Oklahoma Department of Civil Emergency Management. Interview by author, June 23, 2003, Oklahoma City. Tape recording.
Auf der Heide, Erik. *Disaster Response: Principles of Preparation and Coordination.* St. Louis, MO: C. V. Mosby Company, 1989.
Barbera, Joseph, Anthony Macintyre, and Craig DeAtley. "Ambulances to Nowhere: America's Critical Shortfall in Medical Preparedness for Catastrophic Terrorism." In *Countering Terrorism: Dimensions of Preparedness*, ed. Arnold Howitt and Robyn Pangi. Cambridge, MA: MIT Press, 2003.
BBC News. "London Blasts: Emergency Advice," *BBCNewsOnline*. July 7, 2005.
Birkland, Thomas. *Lessons of Disaster: Policy Change after Catastrophic Events.* Washington, DC: Georgetown University Press, 2007.
Blakenely, Ray, Director of Operations, Oklahoma Medical Examiner's Office. Interview by author, June 24, 2003, Oklahoma City. Phone interview.
Boin, Arjen and Paul 't Hart. "Public Leadership in Times of Crisis: Mission Impossible," *Public Administration Review* 63, 5 (1998): 544–553.
Buchanan, Rick, Press Secretary to Governor Keating, State of Oklahoma. Interview by author, July 14, 2003. Phone interview.
Buck, George. *Preparing for Terrorism: An Emergency Services Guide.* Albany, NY: Delmar/Thompson Learning, 2002.
Bullard, Joevan, Assistant City Manager, City of Oklahoma City. Interview by author, July 27, 2003, Oklahoma City. Tape recording.
Bunch, Kenneth, Assistant Chief, Oklahoma City Fire Department. Interview by author, June 24, 2003. Tape recording.
Burkle, Ann, Clinical Coordinator, St. Anthony's Hospital Emergency. Interview by author, June 25, 2003, Oklahoma City. Tape recording.
Burstein, Paul. "Policy Domains: Organization, Culture, and Policy Outcomes," *Annual Review of Sociology* 17 (1991): 327–350.
Carroll, John. "Emergency Management on a Grand Scale: A Bureaucrat's Analysis." In *Handbook of Crisis and Emergency Management*, ed. Ali Farazmand. New York: Marcel Dekker, Inc., 2001.
Cassell, Mark. *How Governments Privatize: The Politics of Divestment in the United States and Germany.* Washington, DC: Georgetown University Press, 2002.
Chicago Council on Foreign Relations. *American Public Opinion and Foreign Policy, 2002.* Chicago, IL: Chicago Council on Foreign Relations, 2002.

Citty, Bill, Captain, Public Information Officer, Oklahoma City Police Department. Interview by author, June 26, 2003, Oklahoma City. Tape recording.

City of Oklahoma City. *Emergency Management Operations Plan*. 1994.

Clark, John, Lieutenant, Office of Emergency Management, Oklahoma City Police Department. Interview by author, July 1, 2003. Phone interview.

Clark, Tony. "Nichols Gets Life for Oklahoma Bombing." *CNN.com*. June 4, 1998, via http://www.cnn.com/US/9703/okc.trial/nichols.sentence/ on November 20, 2008.

Clarke, Lee. "Panic: Myth or Reality?" *Contexts* 1, 3 (2002): 21–26.

Cohen, Steven, William Eimicke, and Jessica Horan. "Catastrophe and the Public Service: A Case Study of the Government Response to the Destruction of the World Trade Center," *Public Administration Review* 62, Special Issue (2002).

Coleman, Sally, Jeffrey Brudney, and J. Edward Kellough. "Bureaucracy as a Representative Institution: Toward a Reconciliation of Bureaucratic Government and Democratic Theory," *American Journal of Political Science* 42, 3 (1998): 717–744.

Comfort, Louise. "Cities at Risk: Hurricane Katrina and the Drowning of New Orleans," *Urban Affairs Review* 41, 4 (March 2006): 501–516.

———. "Designing Policy for Action: The Emergency Management System." In *Managing Disaster: Strategies and Policy Perspectives*, ed. Louise Comfort. Durham, NC: Duke University Press, 1988.

———. "Managing Intergovernmental Responses to Terrorism and Other Extreme Events," *Publius: The Journal of Federalism* 32, 4 (Fall 2002): 29–49.

Comfort, Louise, Yesim Sungu, David Johnson, and Mark Dunn. "Complex Systems in Crisis: Anticipation and Resilience in Dynamic Environments," *Journal of Contingencies and Crisis Management* 9, 3 (September 2001): 144–158.

Cortright, Adjutant General Stephen. Writing in Response to an After-Action Report request from the City of Oklahoma City Documentation Team. 1995.

Davidson-Smith, G. "Counterterrorism Contingency Planning and Incident Management," *Domestic Responses to International Terrorism*. Ardsley-on-Hudson, NY: Transnational Publishers, Inc., 1991.

Davis, Gary, Commander, Emergency Medical Service, Oklahoma City Fire Department. Interview by author, June 27, 2003, Oklahoma City. Tape recording.

Dawes, Sharon, Anthony Cresswell, and Bruce Cahan. "Learning from Crisis: Lessons in Human and Information Infrastructure from the World Trade Center Response." Unpublished Manuscript based on a National Science Foundation study, 2002.

Dawes, Sharon, Thomas Birkland, Giri Kumar Tayi, and Carrie Schneider. *Information, Technology, and Coordination: Lessons from the World Trade Center Response*. Albany, NY: Center for Technology in Government, 2004.

Derthick, Martha. *Policymaking for Social Security*. Washington, DC: Brookings Institution, 1979.

Drabek, Thomas. *Disaster in Aisle 13: A Case Study of the Coliseum Explosion at the Indiana State Fairgrounds, October 31, 1963*. Columbus, OH: Ohio State University, 1968.

Drabek, Thomas and Gerard Hoetmer. *Emergency Management: Principles and Practice for Local Government*. Washington, DC: International City Management Association, 1991.

Dynes, Russell. "Interorganizational Relations in Communities under Stress." In *Disasters: Theory and Research*, ed. E. L. Quarantelli. Thousand Oaks, CA: Sage Publications, 1978.

———. *Organized Behavior in Disaster*. Lexington, KY: Heath Lexington Books, 1970.

Dynes, Russell and E. L. Quarantelli. "Property Norms and Looting: Their Patterns in Community Crisis," *Phylon* 31 (1960): 168–182.

Emergency Medical Services Authority. *Terror in the Heartland: The EMSA Response*. 1995.

Falkenrath, Richard. "The Problems of Preparedness: Challenges Facing the U.S. Domestic Preparedness Program." BSCIA Discussion Paper 2000-28, ESDP Discussion Paper ESDP-2000-05, John F. Kennedy school of Government, Harvard University.

Farazmand, Ali (ed.). *Handbook of Crisis and Emergency Management*. New York: Marcel Dekker, Inc., 2001.

Farha, Dr. Bryan, Counselor and Professor, Oklahoma State University. Interview by author, June 24, 2003, Oklahoma City. Tape recording.

Federal Emergency Management Agency. *CONPLAN: United States Government Interagency Domestic Terrorism Concept of Operations Plan.* Washington, DC: Federal Emergency Management Agency, 2001.

———. *Consequence Management for Nuclear, Biological and Chemical Terrorism: The Federal Response Plan.* Washington, DC: Federal Emergency Management Agency, 1996.

———. *Federal Response Plan.* Washington, DC: Federal Emergency Management Agency, 2003.

———. *Guide for the Development of State and Local Emergency Operations Plans.* Washington, DC: Federal Emergency Management Agency, 1990.

———. *Managing the Emergency Consequences of Terrorist Incidents: Interim Planning Guide for State and Local Governments.* Washington, DC: Federal Emergency Management Agency, 2002.

———. *National Incident Management System.* Washington, DC: Federal Emergency Management Agency, 2005.

Federal Emergency Management Agency, Emergency Management Institute. *Independent Study Course: ICS Incident Command System.* Washington, DC: Federal Emergency Management Agency, 1998.

Ferreira, Steve, Non-Commissioned Officer in Charge of Disaster Management, Tinker Air Force Base. Interview by author, June 21, 2004, Tinker Air Force Base. Phone interview.

Fesler, James and Donald Kettl. *The Politics of the Administrative Process.* Chatham, NJ: Chatham House Publishers Inc., 1996.

Flin, Rhona and Kevin Arbuthnot. *Incident Command: Tales from the Hotseat.* Burlington, VT: Ashgate Publishing Company, 2002.

Form, William and Sigmund Nosow. *Community in Disaster.* New York: Harper & Brothers Publishers, 1958.

Freedberg, Sydney. "Homeland Defense Effort Breaks Down Walls of Government," *GovExec.com Daily Briefing.* October 19, 2001.

Frucht, Leora. "Canada to Learn Lessons from Israel in Improving Emergency Response," *JewishToronto.net.* downloaded on August 22, 2005.

Goggin, Malcolm, Ann Bowman, James Lester, and Laurence O'Toole. *Implementation Theory and Practice: Toward a Third Generation.* New York: Harper Collins Publishers, 1988.

Gonzales, Sam, Chief, Oklahoma City Police Department. Interview by author, July 1, 2003, Oklahoma City. Phone interview.

Goodsell, Charles. *The Case for Bureaucracy: A Public Administration Polemic.* Chatham, NJ: Chatham House Publishers, 1985.

Grimes, Mike, Captain, Oklahoma Highway Patrol. Interview by author, June 23, 2003, Oklahoma City. Tape recording.

Hale, Sue, Assistant Managing Editor, *Daily Oklahoman.* Interview by author, June 25, 2003, Oklahoma City. Tape recording.

Hale, Troy, Plans, Operations, and Support, Oklahoma National Guard. Interview by author, June 13, 2005. Phone interview.

Hampton, Debby, Local Volunteer Manager, American Red Cross. Interview by author, June 23, 2003, Oklahoma City. Tape recording.

Hansen, Jon, Assistant Chief, Public Information Officer, Oklahoma City Fire Department. Interview by author, June 23, 2003, Oklahoma City. Tape recording.

Hansen, Jon. "Working with the Media," *Fire Engineering: Special Issue: Oklahoma City Bombing, Volume 1.* October 1995.

Heclo, Hugh. "Issue Networks and the Executive Establishment." In *The New American Political System,* ed. Anthony King. Washington, DC: American Enterprise Institute, 1978.

Hightower, Henry and Michel Coutu. "Coordinating Emergency Management: A Canadian Example." In *Disaster Management in the U.S. and Canada,* ed. Richard Sylves and William Waugh. Springfield, IL: Charles C. Thomas, 1996.

Hill, Ed, Captain, Emergency Response Team, Oklahoma City Police Department. Interview by author, June 24, 2003, Oklahoma City. Tape recording.

Hill, Mimi. "Government Training Citizens in Civil Defense," *USA Today*. February 13, 2003.

Hinmin, Eve. "Explosion and Collapse: Disaster in Milliseconds," *Fire Engineering: Special Issue: Oklahoma City Bombing, Volume 1*. October 1995.

Howitt, Arnold and Robin Pangi. "Intergovernmental Challenges of Combating Terrorism." In *Countering Terrorism: Dimensions of Preparedness*, ed. Arnold Howitt and Robin Pangi. Cambridge, MA: MIT Press, 2003, 37–57.

Hsu, Spencer. "After the Storm, Chertoff Vows to Reshape DHS," *The Washington Post*. November 14, 2005.

Injury Prevention Service Oklahoma State Department of Health. *Investigation of Physical Injuries Directly Associated with the Oklahoma City Bombing*. Downloaded from www.health.state.ok.us/program/injury/okcbom.html on May 10, 2003.

Jenkins, Brian. "The Future Course of International Terrorism," *The Futurist*. July/August 1987.

———. "Will Terrorists Go Nuclear?" *Orbis* 3 (Autumn 1985): 507–516.

Jennings, Edward and Jo Ann Ewalt. "Interorganizational Coordination, Administrative Consolidation, and Policy Performance," *Public Administration Review* 58, 5 (1998): 417–428.

Jordan, Lara Jakes. "Former FEMA Director Blames State, Local Officials for Response," *State & Local Wire*. September 27, 2005.

Kettl, Donald. "Contingent Coordination: Practical and Theoretical Puzzles for Homeland Security," *American Review of Public Administration* 33, 3 (2003): 253–277.

———. *The Regulation of American Federalism*. Baton Rouge, LA: Louisiana State University Press, 1983.

Kickert, Walter, Erik-Hans Klijn and Joop Koppenjan. "Introduction: A Management Perspective on Policy Networks." In *Managing Complex Networks*, ed. Klijn Kickert and Koppenjan. London: Sage Publications, 1997.

Kingdon, John. *Agendas, Alternatives, and Public Policy*. New York: Harper Collins, 1995.

Kirschenbaum, Alan. *Chaos Organization and Disaster Management*. New York: Marcel Dekker, Inc., 2004.

Kreps, Gary. "The Organization of Disaster Response: Some Fundamental Theoretical Issues." In *Disasters: Theory and Research*, ed. E. L. Quarantelli. Thousand Oaks, CA: Sage Publications, 1978, 65–85.

Laquer, Walter. *The Age of Terrorism*. Boston, MA: Little, Brown, 1987.

Lewis, Ralph. "Management Issues in Emergency Response." In *Managing Disaster: Strategies and Policy Perspectives*, ed. Louise Comfort. Durham, NC: Duke University Press, 1988, 163–179.

Lin, Nan. "Building a Network Theory of Social Capital." In *Social Capital: Theory and Research*, ed. Nan Lin, Karen Cook, and Ronald Burt. New York: Aldine De Gruyter, 2001.

———. *Social Capital: A Theory of Social Structure and Action*. Cambridge: Cambridge University Press, 2001.

Lipsky, Michael. *Street-Level Bureaucracy: Dilemmas of the Individual in Public Services*. New York: Russel Sage Foundation, 1980.

MacRae, Duncan and Dale Whittington. *Expert Advice for Policy Choice: Analysis and Discourse*. Washington, DC: Georgetown University Press, 1997.

Maniscalco, Paul and Hank Christen. *Understanding Terrorism and Managing the Consequences*. Upper Saddle River, NJ: Prentice Hall, 2001.

Marrs, Gary, Chief, Oklahoma City Fire Department. Interview by author, June 24, 2003, Oklahoma City. Tape recording.

Marrs, Gary. "Report from Fire Chief," *Fire Engineering: Special Issue: Oklahoma City Bombing, Volume 1*. October 1995.

May, Peter and W. Williams. *Disaster Policy Implementation: Managing Programs under Shared Governance*. New York: Plenum Press, 1986.

Mayer-Schonberger, Viktor. "Emergency Communications: The Quest for Interoperability in the United States and Europe." In *Countering Terrorism: Dimensions of Preparedness*, ed. Arnold Howitt and Robyn Pangi. Cambridge, MA: MIT Press, 2003.

Mazmanian, Daniel and Paul Sabatier. *Implementation and Public Policy*. Glenview, IL: Scott, Foresman, and Co., 1983.

Moss, Ron, Rescue Operations Commander, Oklahoma City Fire Department. Interview by author, June 24, 2003, Oklahoma City. Tape recording.

Moyser, George. "Non-Standardized Interviewing in Elite Research." In *Studies in Qualitative Methodology*, ed. Robert Burgess. Greenwich, CT: JAI Press, 1988.

Murphy, Mike, Commander, Emergency Medical Services Authority. Interview by author, June 26, 2003, Oklahoma City. Tape recording.

Nakamara, Akira. "Preparing for the Inevitable: Japan's Ongoing Search for Best Crisis Management Practices." In *Managing Crises: Threats, Dilemmas, Opportunities*, ed. Uriel Rosenthal, R. Arjen Boin, and Louise Comfort. Springfield, IL: Charles C. Thomas Publisher, Ltd., 2001.

National Commission on Terrorist Attacks upon the United States. *The 9/11 Commission Report*. New York: W. W. Norton, 2003.

Nice, David and Ashley Grosse. "Crisis Policy Making: Some Implications for Program Management." In *Handbook of Crisis and Emergency Management*, ed. Ali Farazmand. New York: Marcel Dekker, Inc., 2001.

Norick, Ron, Mayor, City of Oklahoma City. Interview by author, June 23, 2003, Oklahoma City. Tape recording.

Nunn-Lugar-Domenici Act. Title XIV of the National Defense Authorization Act of 1996. Public Law 104-201, 1996.

Oklahoma City Department of Civil Emergency Management. *After-Action Report: Alfred P. Murrah Federal Building Bombing: Detailed Summary of Daily Activity*.

Oklahoma City Document Management Team. *The City of Oklahoma City Alfred P. Murrah Federal Building Bombing April 19, 1995 Final Report*. Stillwater, OK: Fire Protection Publications, 1996.

———. Interview with Gary Marrs, Incident Commander and Chief, Oklahoma City Fire Department. Transcription. Oklahoma City National Memorial Archives, October 7, 1999.

———. Interview with Mike Murphy, Command, EMSA. Transcription. Oklahoma City National Memorial Archives, March 21, 2000.

———. Interview with Mike Shannon, Oklahoma City Fire Department. Transcription. Oklahoma City National Memorial Archives, undated.

———. Interview with Ray Blakenely, Director of Operations, Oklahoma Medical Examiner's Office. Transcription. Oklahoma City National Memorial Archives, October 20, 1999.

———. Interview with Ron Norick, Mayor, City of Oklahoma City. Transcription. Oklahoma City National Memorial Archives, n.d.

Oklahoma City National Memorial Institute for the Prevention of Terrorism. *Oklahoma City, 7 Years Later, Lessons for Other Communities*. 2002.

Oklahoma City Police Department. *After-Action Report: Alfred P. Murrah Federal Building Bombing Incident, April 19, 1995*. 1995.

Omnibus Diplomatic Security Act of 1986. Public Law No: 99-399.

Ornstein, Norman. "Some Steps Congress Can Take to Prevent Another Katrina," *Congress Inside Out*. September 14, 2005.

O'Toole, Laurence. "Multiorganizational Implementation: Comparative Analysis for Wastewater Treatment." In *Strategies for Managing Intergovernmental Policies and Networks*, ed. Robert Gage and Myrna Mandell. New York: Praeger, 1989.

Pangi, Robyn. "Consequence Management in the 1995 Sarin Attacks on the Japanese Subway System." In *Countering Terrorism: Dimensions of Preparedness*, ed. Arnold Howitt and Robyn Pangi. Cambridge, MA: MIT Press, 2003.

Penn, Billy, Public Affairs Officer, Federal Emergency Management Agency. Interview by author, July 12, 2003. Phone interview.
Perry, Ronald. "Incident Management Systems in Disaster Management," *Disaster Prevention and Management* 12, 5 (2003): 405–412.
Peters, B. Guy. "The Problem of Bureaucratic Government," *The Journal of Politics* 43, 1 (1981): 56–82.
Phillips, Brenda. "Qualitative Methods in Disaster Research." In *Methods of Disaster Research*, ed. Robert Stallings. Philadelphia, PA: Xlibris Corporation, 2002.
Porter, D. "Federalism, Revenue Sharing, and Local Government." In *Public Policy-Making in the Federal System*, ed. Charles Jones and Robert Thomas. Thousand Oaks, CA: Sage Publications, 1976.
Posner, Paul. *Combating Terrorism: Intergovernmental Partnership in a National Strategy to Enhance State and Local Preparedness*. Testimony before the U.S. House, Committee on Government Reform, Subcommittee on Government Efficiency, Financial Management, and Intergovernmental Relations. Washington, DC: General Accounting Office, 2002.
Pressman, Jeffrey and Aaron Wildavsky. *Implementation*. Los Angeles, CA: University of California Press, 1984.
Proctor, Chief, Tinker Air Force Base Fire Department. Interview by author, July 14, 2004, Tinker Air Force Base. Phone interview.
Putnam, Robert and Kristin Goss. "Introduction." In *Democracies in Flux: The Evolution of Social Capital in Contemporary Society*, ed. Robert Putnam. Oxford: Oxford University Press, 2002.
Quarantelli, E. L. "The Controversy on the Mental Health Consequences of Disasters." In *Groups and Organizations in War, Disaster, and Trauma*, ed. R. Ursano. Bethesda, MD: Uniformed Services University of the Health Sciences.
———. (ed). *Disasters, Theory and Research*. Thousand Oaks, CA: Sage Publications, 1978.
Reilly, John. *American Public Opinion and U.S. Foreign Policy, 1999*. Chicago, IL: Chicago Council on Foreign Relations, 1999.
Ricks, Bob, Special Agent in Charge, Federal Bureau of Investigation. Interview by author, July 8, 2003. Phone interview.
Riley, K. Jack and Bruce Hoffman. *Domestic Terrorism: A National Assessment of State and Local Preparedness*. Santa Monica, CA: RAND Corporation, 1995.
Ripley, Randall and Grace Franklin. *Policy Implementation and Bureaucracy*. Chicago, IL: Dorsey Press, 1982.
Robert T. Stafford Disaster Relief and Emergency Assistance Act, as amended by Public Law 106-390, October 30, 2000. USC 42, Chapter 68. [As amended by Pub. L. 103-181, Pub. L. 103-337, and Pub. L. 106-390] (Pub. L. 106-390, October 30, 2000, 113 Stat. 1552-1575).
Robinson, Mark. "EMS Treatment and Transport," *Fire Engineering: Special Issue: Oklahoma City Bombing, Volume 1*. October 1995.
Rodriguez, Havidan, Enrico Qarantelli, and Russell Dynes (eds). *Handbook of Disaster Research*. New York: Springer, 2007.
Rosenthal, Uriel, Arjen Boin, and Louise Comfort. "The Changing World of Crises and Crisis Management," *Managing Crises: Threats, Dilemmas, and Opportunities*, ed. Rosenthal, Boin, and Comfort. Springfield, IL: Charles C. Thomas Publisher, Ltd., 2001.
Rosenthal, Uriel, Michael Charles, and Paul Hart. *Coping with Crises: Management of Disasters, Riots, and Terrorism*. Springfield, IL: Charles Thomas Publisher, 1989.
Ross, G. Alexander. "Organizational Innovation in Disaster Settings." In *Disasters: Theory and Research*, ed. E. L. Quarantelli. Thousand Oaks, CA: Sage Publications, 1978, 215–232.
Rudman, Warren, Richard Clarke, and Jamie Metzl. *Emergency Responders: Drastically Underfunded, Dangerously Underprepared*. Report of an Independent Task Force Sponsored by the Council on Foreign Relations, 2003.
Schein, Edgar. *Organizational Culture and Leadership*. Indianapolis, IN: Jossey-Bass, 2004.
Schneider, Saundra. *Flirting with Disaster: Public Management in Crisis Situations*. New York: M.E. Sharpe, 1995.

Shannon, Mike, Rescue Operations Chief, Oklahoma City Fire Department. Interview by author, June 26, 2003, Oklahoma City. Tape recording.

———. "Rescue Operations: Doing Battle with the Building," *Fire Engineering: Special Issue: Oklahoma City Bombing, Volume 1*. October 1995.

Smithson, Amy and Levy, Leslie-Anne. *Ataxia: Chemical and Biological Terrorism Threat and the U.S. Response*. Washington, DC: Stimson Center, 2000.

Stake, Robert. "Case Studies." In *Handbook of Qualitative Research*, ed. Norman Denzin and Yvonna Lincoln. Thousand Oaks, CA: Sage Publications, 1994.

Stillman, Richard. *The American Bureaucracy: The Core of Modern Government*. Chicago, IL: Nelson-Hall Publishers, 1996.

Stockton, Dan, Lieutenant, Public Information Officer, Oklahoma Highway Patrol. Interview by author, June 27, 2003, Oklahoma City. Tape recording.

Storment, Steve, Chief, Arizona USAR Team for FEMA. Interview by author, June 25, 2003, Arizona. Phone interview.

Thompson, James. *Rolling Thunder: Understanding Policy and Program Failure*. Chapel Hill, NC: University of North Carolina Press, 1980.

Thurmaier, Kurt and Curtis Wood. "Interlocal Agreements as Overlapping Social Networks: Picket-Fence Regionalism in Metropolitan Kansas City," *Public Administration Review* 62, 5 (2002): 585–598.

Toft, Brian and Simon Reynolds. *Learning from Disasters*. Oxford: Butterworth-Heinemann, Ltd., 1994.

Torr, James. *Responding to Attack: Firefighters and Police*. San Diego, CA: Lucent Books, 2004.

U.S. Commission on National Security in the 21st Century. *New World Coming: American Security in the 21st Century*. Washington, DC: Government Printing Office, 1998.

U.S. Department of Homeland Security. *National Response Plan*. Washington, DC: Government Printing Office, 2004.

U.S. General Accounting Office. *Combating Terrorism: Federal Trends and Implications for First Responders*. Washington, DC: Government Printing Office, 2001.

———. *Combating Terrorism: FEMA Continues to Make Progress in Coordinating Preparedness and Response*. Washington, DC: Government Printing Office, 2001.

———. *Combating Terrorism: Linking Threats to Strategies and Resources*. Washington, DC: Government Printing Office, 2000.

———. *Combating Terrorism: Opportunities to Improve Domestic Preparedness Program Focus and Efficiency*. Washington, DC: Government Printing Office, 1998.

———. *Combating Terrorism: Selected Challenges and Related Recommendations*. Washington, DC: Government Printing Office, 2001.

———. *Disaster Management: Improving the Nation's Response to Catastrophic Disasters*. Washington, DC: Government Printing Office, 1993.

———. *Homeland Security: Progress Made; More Direction and Partnership Sought*. Washington, DC: Government Printing Office, 2002.

———. *Major Management Challenges and Program Risks: Federal Emergency Management Agency*. Washington, DC: Government Printing Office, 2003.

Wagman, David. "Emergency Management and Civil Defense." In *Homeland Security: Best Practices for Local Government*, ed. Roger Kemp. Washington, DC: ICMA, 2003.

Warren, Ronny, Patrol Sergeant, Emergency Medical Technician, ERT, Oklahoma City Police Department. Interview by author, June 20, 2003. Phone interview.

Waugh, William. "Co-ordination or Control: Organizational Design and the Emergency Management Function," *Disaster Prevention and Management: An International Journal* 2, 4 (1993): 17–31.

———. "Current Policy and Implementation Issues in Disaster Preparedness." In *Managing Disaster: Strategies and Policy Perspectives*, ed. Louise Comfort. Durham, NC: Duke University Press, 1988.

———. "Managing Terrorism as an Environmental Hazard," *Handbook of Crisis and Emergency Management*. New York: Marcel Dekker, Inc., 2001.

———. "Regionalizing Emergency Management: Counties as State and Local Government," *Public Administration Review* 54, 3 (1994): 253–258.

———. *Terrorism and Emergency Management: Policy and Administration*. New York: Marcel Dekker, Inc., 1990.

Waugh, William and Richard Sylves. "Organizing the War on Terrorism," *Public Administration Review* 62, Special Issue (2002): 145–153.

Weber, Max. "Bureaucracy." In *Critical Studies in Organization and Bureaucracy*, ed. Frank Fisher and Carmen Sirianni. Philadelphia, PA: Temple University Press, 1984.

White, Leonard. *Introduction to the Study of Public Administration*. New York: Macmillan, 1939.

Wiarda, Howard. *Introduction to Comparative Politics: Concepts and Processes*. Ft. Worth, TX: Harcourt College Publishers, 2000.

Wilson, James Q. *Bureaucracy: What Government Agencies Do and Why They Do It*. Philadelphia, PA: Basic Books, 1989.

Wise, Charles and Rania Nader. "Organizing the Federal System for Homeland Security: Problems, Issues, and Dilemmas," *Public Administration Review* 62, S1 (2002): 44–57.

Yates, Steven. "The Oklahoma City Bombing: A Morass of Unanswered Questions." *LewRockwell.com*. May 19, 2001 via http://www.lewrockwell.com/yates/yates33.html on August 21, 2002.

Yin, R. K. *Case Study Research Design Methods*. Thousand Oaks, CA: Sage Publications, 2003.

Young, Cornelius, Major, Oklahoma City Fire Department. Interview by author, June 23, 2003, Oklahoma City. Tape recording.

Index

9/11 14, 91–3

after action report (AAR) 14, 43
American Red Cross 29–30, 67, 70, 87
Ashwood, Albert, Director, ODCEM 37, 40, 58, 67–8, 90
Auf der Heide, Erik 3, 7, 38

Birkland, Thomas 3, 7
Blakenely, Ray Director of Operations, Medical Examiner's Office 44, 78
Bullard, Joevan Assistant City Manager 38–9, 61, 70, 72, 74, 89
Bunch, Kenneth, Assistant Chief, OCFD 36, 39, 54, 59–60
Burkle, Ann Clinical Coordinator, St. Anthony Hospital Emergency Department 35, 37, 45, 60, 71–2, 76, 93

CERT *see* community emergency response teams
CISD *see* critical incident stress debriefings
Citty, Bill, Information Officer, Captain, OCPD 36, 53, 57, 72
Clark, John, Lieutenant, OCPD Office of Emergency Management 42, 55, 68, 73–4
Comfort, Louise 5, 14, 34, 55
community emergency response teams 91–2
Cortright, Stephen Adjutant General Oklahoma National Guard 59, 70
critical incident stress debriefings 77, 84

Davis, Gary, Commander, EMS for the OCFD 36, 42, 60, 73, 75
Department of Homeland Security (DHS) 84, 89, 92
disaster mortuary teams (D-MORT) 27, 78
donations 6, 41–3, 84, 97
Dynes, Russell 41, 53, 74

Emergency Management Institute (EMI) 61
Emergency Medical Service Authority (EMSA) 20, 28–30, 38, 55, 61, 66

emergency medical system (EMS) 20, 45, 75, 84
Emergency Operations Center (EOC) 86, 88
emergency support functions (ESF) 12, 15, 66, 86–8
Emmitsburg 51, 59, 61, 84, 91

family assistance center (FAC) 30, 46, 67, 78–9, 87, 89
Federal Bureau of Investigation (FBI) 24, 29, 45, 66, 84
 crime scene and rescue scene 6, 57
 role 26, 78, 85–6
Federal Emergency Management Agency (FEMA) 55
 crime scene and rescue scene 6, 57
 response plans and guidelines 12, 39–40, 66–8, 72, 85, 89, 98
 role 27–30, 84
Ferreira, Steve Non-Commissioned Officer in Charge for Disaster Management, TAFB 38, 70
Form, William 53
Freedberg, Sydney 39

Gonzales, Sam, Chief, OCPD 26, 44, 58–9, 61, 68–9, 73
Goodsell, Charles 10
Grimes, Mike, Captain, Oklahoma Highway Patrol 44, 62, 67, 71
 crime scene and rescue 40, 57
 donations management 43, 74

Hale, Sue, Assistant Managing Editor, *Daily Oklahoman* 59, 73, 77, 92
Hale, Troy, Plans, Operation, and Support, Oklahoma National Guard 38
Hampton, Debby, Local Volunteer Coordinator, American Red Cross 42, 59, 77–8
Hansen, Jon, PIO OCFD 38, 57, 59, 72–3, 75, 88
Heclo, Hugh 9

Hill, Ed, Emergency Response Team Captain, OCPD 35, 38, 53, 60–2, 89
Howitt, Arnold 53

Incident Command System (ICS) 55, 57–8, 61–3,
  description 36, 54, 89, 95
  implementation at incident 66, 68, 84, 86, 91
  incident commander 54–5, 58, 92, 95

Kettl, Donald 53

Lin, Nan 58
Lipsky, Michael 9–10

Marrs, Gary, Incident Commander and Chief, OCFD 23, 45, 77
  command 26, 35–6, 57–9,
  crime scene and rescue 40, 69
  Emmitsburg training 61, 67, 85
Moss, Ron, Rescue Operations Commander, OCFD 41, 55
Murphy, Mike, Operations Officer, EMSA 61, 68, 75
  communications 37, 55, 71
  triage 45
mutual aid agreements 34, 59, 97
  implementation 20, 69, 84, 87–8, 93

National Incident Management System 39, 85, 89–90
National Response Framework 4, 85, 89, 91
National Response Plan 4, 91
network 51, 53, 57–62, 96–8
  development 85, 93–4
  impact on response 5–6, 39, 84
  theoretical issues 8–9
Nichols, Terry 21
NIMS *see* National Incident Management System
Norick, Ron, Mayor, City of Oklahoma City 28, 38, 70, 72
Nosow, Sigmund 53, 85, 89–90

NRF *see* National Response Framework
NRP *see* National Response Plan

Oklahoma Department of Civil Emergency Management (ODCEM) 28, 55, 69

Pangi, Robin 53
Penn, William, Public Affairs Officer, FEMA 39, 54, 57, 71, 83
perimeter 36, 78
  challenges 6, 41–2, 73–4
  establishment 26–7, 29–30, 43–4, 59, 66–7
  importance 92, 95–6
Peters, B. Guy 8
Phillips, Brenda 12
Pressman, Jeffrey 8
public information officer 29, 38, 72, 86, 88, 93
Putnam, Robert 58

Quarantelli, E. L. 60, 74

Ricks, Bob Special Agent in Charge, FBI 26, 57–9, 62, 68–9

Salvation Army 29, 67, 84, 87
Schein, Edgar 10
Shannon, Mike, Rescue Operation Chief, OCFD 21, 35, 55, 58–60
social capital 51, 58–60
Stockton, Dan, PIO, Lieutenant, Oklahoma Highway Patrol 35–6, 59, 72–3
Sylves, Richard 51

TAFB *see* Tinker Air Force Base
Thompson, Ken 79
Tinker Air Force Base 24, 77
  network and training 59
  role in response 29–31, 67, 69–70
training 33, 34, 46, 58–63, 68, 84–5, 89, 96, 98
  bureaucratic culture building 10, 51, 66
  lack of 6, 23, 33, 44
  network building 5, 91–4

triage 66, 75–6, 93, 96
  implementation 5, 26–8
  importance 45, 95
  multiple triage stations 41, 71

Urban Search and Rescue (USAR) 27–8, 67, 88–9

volunteers 6, 41–3, 66–7, 84, 91–2, 95, 97

Warren, Ronny, Patrol Sergeant and Emergency Medical Technician, OCPD 55, 57
Waugh, William 33, 37, 51, 67
Wildavsky, Aaron 8
Wilson, James Q. 8–9, 52

Young, Cornelius, Major, OCFD 60

www.ingramcontent.com/pod-product-compliance
Lightning Source LLC
Chambersburg PA
CBHW070335230426

43663CB00011B/2330